John Batchelor
Ian Hogg

Artillery

CHARLES SCRIBNER'S SONS • New York

Library of Congress number 72-6654

SBN 684-13092-0

Published by arrangement with
Ballantine Books Inc.
201 East 50th Street, New York, N.Y. 10022

Printed in USA by
Von Hoffmann Press, Inc.

Layout and make-up Mike Jarvis

2 **Mohammed's Great Gun** which was the grandfather of all screw-guns. It came to pieces to simplify both manufacture and transport, and the slotted borders at each end allowed the pieces to be moved by handspikes. Of 25 inch calibre it weighed 19 tons and required 60 oxen to draw it.

3

3 **The Great Gun of Moscow.** No means of elevation are apparent but this was probably unnecessary, as the monster was solely for battering at stone defences at point-blank range.

4 **A Burgundian 'Falcon'** of the 15th Century. An early and interesting method of achieving variable elevation by using a pivoted cradle to carry the gun and anchoring it on the elevation arcs at the rear.

4

The gun

The origins of gunpowder and the gun are hidden away in what the lawyers call "Time Immemorial". The only certain thing is that gunpowder came first, followed at a respectful distance by the gun. This sounds a little odd, but the oddity is compounded by the fact that the British persist in calling the explosive substance "Gunpowder" – which suggests that it was produced in order to make something of the gun – whereas everybody else calls it, rather more logically, "Black Powder", keeping it free from any untoward alliance with ordnance. The late Lt-Colonel H. W. L. Hime of the Royal Artillery, an able historian and painstaking scholar, spent the latter years of the nineteenth and early years of the twentieth centuries delving into libraries all over Europe and bit by bit building up what information there was to be gained on the early history of black powder. He refuted fairly conclusively the oft-repeated tale that it was invented by the Chinese and exported to the West, and also showed that the fabled "Greek Fire" was in no way related to black powder. In addition, he broke the anagram with which Roger Bacon, philosopher and monk, had concealed his formula for powder in *De Secretis Operibus Artis et Naturae,* a treatise on contemporary scientific beliefs written in about 1248. But beyond this Hime was not prepared to go, and nor has any other reputable scholar since then. The fact remains that the inventor of gunpowder is unknown to us and never will be known to us; the best that can be said is that the substance appears to have become known to alchemists and similar scientific workers at some time during the first half of the thirteenth century. Moreover, it would appear that its properties were accepted as being purely pyrotechnic at that time, and the substance was probably filled into paper cases to make a species of firework for spectacular effect and the general mystification of the populace, rather than with any warlike intent.

Several stories have been told to the effect that one Berthold Schwarz, a monk of the Black Forest area of Germany, in the course of experiments with black powder managed to fire a charge in an apothecary's mortar, blowing the cover from the vessel in the process and leading Schwarz to design a projectile-throwing weapon. Unfortunately there is no evidence to support this tale, indeed there is no evidence to support the reputed existence of Schwarz at all, and so, as with powder, so with the gun. We cannot praise (or blame) any given individual for the invention which has been so instrumental in shaping the world in the fashion it is today.
Speculation ceases with a document, dated the 11th February 1326, which authorises the Priors, Gonfalonier and twelve good men to appoint persons to superintend the manufacture of brass cannon and iron balls for the defence of Florence. In the same year the Millimete Manuscript, in the possession of the Bodleian Library, Oxford, depicts a cannon in the act of being fired; the vase-like weapon is being touched off with a hot iron and the projectile is a form of arrow. The first projectiles were arrow-like, since such projectiles were known to be successful with bows and other engines, and a wrapping of tow or rag would suffice to hold the bolt more or less central in the gun barrel and also would help to confine the gas behind it and ensure some sort of efficiency of the powder's explosion. It is interesting to see that the Florentine record specifically refers to iron balls being prepared for their cannon, so that balls and arrows or bolts must have co-existed for a while in the earliest days until the ball completely usurped the bolt.

At this time the Genoese and the English were trading together and there is every likelihood that through these channels either the knowledge or some actual specimen of ordnance made its way to England, for 1327 gives the first reputed record of guns in use in England and Scotland. John Barbour, Archdeacon of Aberdeen, later wrote in his *Metrical Life of King Bruce,* dealing with the first campaign of Edward III against the Scots in 1327,

"Twa novelties that day they saw
That forenst in Scotland had been nane
Timbers for helms was the ane
That they brought them of great beautie
And also wonder for to see
The other crakys were of war
That they before heard never air"

5 **The Kufstein Mortar** a 16th Century siege weapon on its wooden bed. Elevation is not adjustable, and ranging had to be done by either moving the mortar about bodily or altering the gunpowder charge.

6 **British Land Service Mortar** of 1769, a bronze barrel, trunnioned at the rear into a wooden bed. Elevation was controlled by wedges. Ordnance of this type was also used by the Navy for shore bombardment.

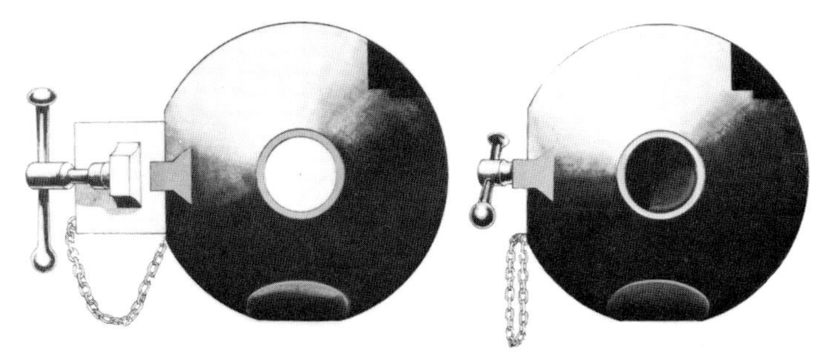

7 **Krupp's Sliding Wedge** in one of its earlier applications. The gun is still massive in comparison with the breech block, for factors of safety were treated liberally. A simple screw gear withdraws the block.

The only drawback to this statement is the fact that Barbour was writing many years after the event. He was born about 1324 and was obviously writing from hearsay, and no corroborative evidence has ever been produced to bear him out. Similar statements by Grafton refer to artillery in 1267 but there is no doubt that Grafton was also dealing with events which occurred before he was born, and the artillery in question was in fact the usual engines of the time.

In 1338 comes the first reliable French reference to ordnance, a document relating to provision made at the Arsenal at Rouen for equipment to be sent to ships fitting out at Harfleur for an attack on Southampton, and it mentions iron cannon provided with 48 feathered bolts of iron, and one pound of saltpetre and half a pound of sulphur for propelling these arrows. These amounts, when combined with charcoal, would have made up about two and a half pounds of black powder of the contemporary blend, and if this was the supply to propel 48 arrows it argues that the gun must have been a relatively small weapon, with the arrows weighing no more than about half a pound each.

The accounts of the Bailiffs of St Omer for 1342 contain a reference to the inventory of the castle of Rihoult in Artois wherein cannon having a separate chamber, held in place by a wedge and fired by a red-hot iron are mentioned. In 1345 the first English reference is seen in the Privy Wardrobe accounts for 1st February, showing that Edward III ordered the Keeper of the Tower Wardrobe to repair ships' guns and ammunition for the King's expedition to France. This, of course, led to the use of cannons at the Battle of Crecy in 1346; but their efficiency is in some doubt, for contemporary references speak disparagingly of their use "to frighten the horses".

On the 20th March 1375 an order was received at Caen from Jehan le Mercier, one of the Counsellors of the King of France, for the manufacture of *"un grand canon de fer"*, and Bernard de Monferrat *"maitre les canons"* was instructed to superintend the operation. Excellent records were kept, and make interesting reading, since this is the earliest known description of cannon-making.

Two days were spent erecting three forges in the market place and surrounding them with a wooden fence to protect the gunsmiths from interruption. Three master smiths and one common smith then commenced receiving pay, and they were provided with eight men to assist, plus one labourer to keep the forges supplied with wood. The process of manufacture occupied six weeks and must have included night work, since there is an item in the account relating to the provision of candles. In the manufacture 2300 lbs of iron were used. The gun tube was built up of longitudinally-placed bars welded together, hooped up by wrought-iron rings. The whole piece was then bound tightly with 90 lbs of rope and covered with cowhide sewn round in order to keep out rain and damp. The vent was covered with an iron flap to keep the powder dry while the gun was loaded. The bed of the cannon was fashioned from a large piece of elm, while more wood secured the barrel to it. At the same time as this gun was made, 24 cannons were fashioned from 421 lbs of *'Nitraille'*, presumably some form of brass or bronze alloy. Unfortunately the records do not give the dimensions of these weapons.

So far the guns seem to have been relatively small; making 24 cannon from 421 lbs of metal gives a weight of but 17 – 18 lbs per gun. But soon the desire to build bigger and better took hold, and by 1377 there is a report of a cannon built for the Duke of Burgundy which was to fire a 450 lb stone shot, giving a probable calibre of 20 inches. But by and large these monster ventures rarely came off, sinch the art of gun-forging was relatively new and the compounding of powder relatively empiric; a mistake in either, in a small calibre would lead to a spectacular failure. But in a large calibre it could lead to a violent fatality for the artillerists, and since these gentlemen were scarce, they preferred not to chance their arm with unusual ordnance – a trait which has persisted ever since.

An interesting concept is revealed by an entry in the Issue Roll, Michaelmas, 8 Richard II (1384); "To Sir Thomas de Beaushamp, Knight., late Captain of Carisbrooke Castle in the Isle of Wight. In monies paid to his own hands for so much monies disbursed by him – viz – to five cannoniers,

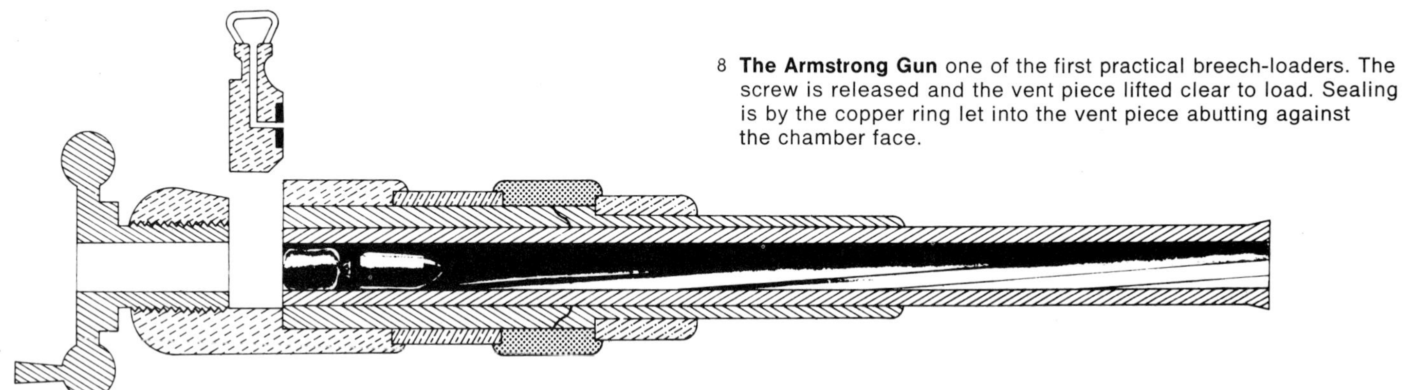

8 **The Armstrong Gun** one of the first practical breech-loaders. The screw is released and the vent piece lifted clear to load. Sealing is by the copper ring let into the vent piece abutting against the chamber face.

each having his own cannon, and for one cannonier with three cannons, for the hire of the same cannoniers and cannons and for powder purchased for the same in the King's service:– £26.5.0d.'' So at this early stage in the cannon's life it was the custom for cannoniers to furnish themselves with their own guns and then hire out to whoever felt the need of them.

The first employment of cannon was undoubtedly for siege purposes; either to make a fearful show outside the castle walls or to actually fling shot at either the castle fabric or the defenders, though the small calibres in use make the efficiency of this employment rather doubtful, But in 1382 came their employment as a field weapon; small cannon called Ribauldequin on two-wheeled carriages placed at the front of the line of battle in order to protect the foot soldiers against sudden attack.

Finally in this review of the earliest days of guns, reference must be made to the first appearance of that keystone, the Master Gunner. The Privy Wardrobe accounts for 6th November 1375 refer to one William Newlyn as the King's Master Gunner at Calais. This appears likely to be the same man mentioned later by Holinshead, who tells that in 1386 two French ships were captured and brought into Sandwich, and on board one of the vessels was found ''a Master Gunner who had led the English at Calais, also divers great engines and guns to beat down walls, also a great quantity of powder''. But he never explained how the poor fellow came to be on a French ship.

From then on, the references abound, and the art of gunnery was off to a good start. The earliest weapons, as we have seen, were built up from hooped bars, whence comes the term ''barrel'' applied to a gun tube, but soon the advantages of casting became apparent. This practice began on the Continent, and the first English brass cannon were cast in 1521 by John Owen. Shortly after this, in 1542, Peter Baud, Ralph Hogge and Peter van Colin, three Sussex ironmasters, successfully began casting iron cannon at Buckstead. By 1574 the business of ordnance was sufficiently involved to necessitate a table being produced to distinguish the types of weapon then in use:

Name	Weight lb	Calibre ins	Shot Weight lb	Charge Weight lb
Robinet	200	1.25	1	0.5
Falconet	500	2	2	2
Falcon	800	2.5	2.5	2.5
Minion	1100	3.25	4.5	4.5
Saker	1500	3.5	5	5
Demi-Culverin	3000	4.5	9	9
Culverin	4000	5.5	18	18
Demi-Cannon	6000	6.5	30	28
Cannon	7000	8	60	44
Extra Cannon	8000	7	42	20
Basilisk	9000	8.75	60	60

One of the most interesting features of this table is the relative weights of shot and powder, which argues that the powder in use at the time was relatively weak.

The provision of metal for such weapons, in the quantities demanded for war, was something of a problem and Grose in his *Military Antiquities* quotes a manuscript of 1578 dealing with the duties of the Master of the Ordnance in

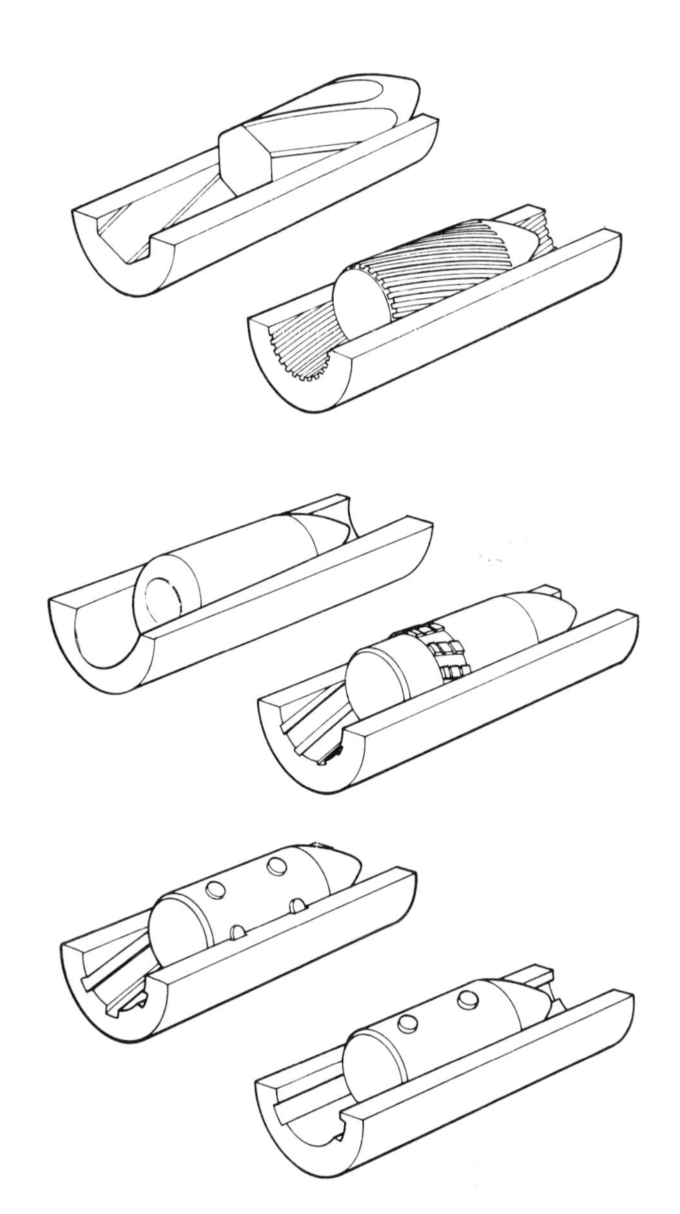

9 **Hexagonal Shell and Bore.** The Whitworth rifling system used this form of hexagonal bore and matching projectile. While theoretically sound, it tends to jam at the pressures and velocities involved in a gun.

10 **Lead Coated Shell** and polygroove rifling. The Armstrong system multiple narrow grooves together with a lead envelope on the shell, which, forced into the grooves, did the job of sealing the propelling gas as well as rotating the projectile.

11 **Oval Bore.** Another contender in the stabilisation field was this oval-sectioned bore and shell designed by Lancaster. While moderately successful in small arms applications, it was less so in artillery, the lack of accuracy being a prime fault.

12 **Vavasseur's Driving Band.** The culmination if many ideas and trials, this became the standard method of imparting rotation to projectiles. The soft copper band acts as a gas seal, spins the shell, and does several other small jobs into the bargain. Vavasseur received £10,000 for his patent.

13 **French Rifling.** The original Rifled Muzzle Loader concept using 6 grooves and 6 rows of studs, attributed to Truille de Beaulieu.

14 **Woolwich Rifling** which was a modification of the French system, using a simpler groove contour and only three grooves. It was selected because of its simpler manufacture and also because it rendered easier the adoption of increasing-twist rifling.

the Field: "As a town is wonne, whether it be by an assault, per force, subtile practise or by other manner given up, be it town, castle, pile, church, bastile or fortress, the Chief Master of the Artillery or his Lieutenant shall ordain that the Master Gunners and their companies shall have the best bell within that place so wonne". From which it appears that the Master Gunners of the day had to be fairly skilled gunfounders as well as artillerists.

At about this time the powder had come under review and had been improved by the technique known as "corning". The original black powder, generally called "Serpentine Powder", was simply made by mixing saltpetre, sulphur and charcoal, roughly in the proportions 60:40:20, though each manufacturer had his own ideas on the subject and his own particular formula. These ingredients were mixed together in the dry state, and the result was a dusty mess which, given the chance and some suitable vibration, would soon separate back into its individual constituents. To counter this, the ingredients were often carried separately and mixed just before being required, so that the powder would be as homogeneous as possible and thus as powerful as possible. The unknown genius who thought up corning solved all these problems in one fell swoop and, into the bargain, made manufacture a somewhat safer business. His invention was the process of mixing the ingredients in the wet state and then spreading the resultant mess thinly on a plate and allowing it to dry. When dried into a solid cake, it was then broken up into grains which could be graded by passing them through a sieve, the dust being removed for incorporation into a later batch. So the powder kept better, being a more secure mixture, and, what was more important, the ignition process was improved, since the flame could pass more readily through the interstices between the grains than through the tightly packed serpentine powder. This in turn led to more certain, more efficient, and more complete ignition of the charge, which led many gunners of the time to believe that "cornpowder" was something quite different and more powerful than the old serpentine. Indeed, so good was corned powder that many elderly pieces of ordnance failed to withstand the increased energy and burst into fragments on being fired.

With this improvement in the quality of powder came the idea to use it as a shell filling as well as a propellant. Until this time the guns had thrown solid projectiles, and any soldier unfortunate enough to get in the way of one was held to be very unlucky indeed. But if it could be made possible to have a hollow ball, and fill it with gunpowder...? By 1588 such shells were in use in the Low Countries, but their operation was by no means attended with complete success, since the explosion of the charge was initiated by the friction of the powder in the shell on impact, an unreliable system and one which led to accidents, for the powder was incapable of distinguishing between the friction due to impact and that generated by the firing of the shell from the cannon, hence premature explosions were the rule rather than the exception. This impasse was not solved until the latter years of the sixteenth century when Sebastian Halle proposed the adoption of a simple time fuze – no more than a length of slow-match inserted into the shell and ignited by the gunner thrusting a linstock down the barrel just before he applied it to the vent. It was later realized that this hazardous game was unnecessary, since the flash of the charge would wash around the shell and ignite the fuze anyway, which made a lot of gunners happy.

Other improvements gradually came on the scene; an elevating screw replaced the quoin or wedge under the breech used to apply elevation to the gun. Master Gunner Nye invented his *"mortar eprouvette"* in 1647, a small mortar firing a specified weight of shot along a graduated scale to test the strength of the powder, a step which made possible the production of some sort of table of charges and ranges. The manufacture of guns was gradually made more precise, together with the manufacture of shot, so that more regular results could be hoped for. For gunnery began as a rule-of-thumb art, precise work being out of the question due to rough construction of the weapons, poor powder of variable quality, and excessive "windage" or clearance between the gun and the projectile when in the barrel. All this, allied to a complete ignorance of the behaviour of the projectile after it had left the muzzle, was the reason that gunnery was almost one of the Black Arts in its early days. But things were slowly improving.

Rifling the gun's barrel to improve the accuracy of the projectile's flight had been proposed several times. Grooved barrels, but with the grooves straight, were developed as early as 1490 but there is no doubt that this was an attempt to overcome powder fouling and not an aid to accuracy. But knowing that rifling was a good thing, and being able to rifle weapons successfully in what was the contemporary equivalent of mass-production was not the same thing. A Dr Lind of Edinburgh produced a rifled one-pounder gun for trial in 1776, but it showed little improvement over an unrifled weapon of the same calibre and no further action was taken. The Royal Artillery Museum at the Rotunda, Woolwich, shows many examples of early rifled weapons, but a great deal had to be done before the theoretical advantages could be turned into practical applications.

We might pause at this point and reflect. From the invention of the gun in – let us say 1325 – to the year 1800 (and later, as will be seen) the advance of artillery had been slow. There was little in the gun made in 1800 which would have puzzled or daunted a Master Gunner from Agincourt or Blenheim. The basic weapon, while it had had slight improvements in manufacture, from its original welded stave form to its cast form, with side-forays into construction from such varied materials as leather and wood, was still a simple smoothbore. Let us now, therefore, examine in detail a typical muzzle-loading cannon of the type current in 1800 to see what is what and how it all works.

The barrel first. Starting at the front we have the muzzle; the thickened circumference around it is the muzzle swell. The body of the barrel is known properly as the "chase", and at intervals there are thicker hoops or "reinforces" put in partly for strength and partly for ornament. At the point of balance are two rounded protrusions, the trunnions, which fit into prepared surfaces on the carriage and allow the gun to be elevated and depressed. Some guns light enough to be lifted off their carriages had handles above the trunnions, generally cast in the shape of ornamental fish and thus usually called the "dolphins". At the rear of the gun the metal is rapidly reduced in diameter and finished off in an ornamental knob or loop; this is the cascable, and generally speaking a solid cascable in-

15 **The Armstrong Field Gun.** Armstrong's 40-pounder of 35 hundredweight, 1862. This "Pattern G" model embodied a number of small improvements on its fore-runners, and 810 of them were built. They were not declared obsolete until 1920.

dicates a field piece, while a pierced cascable indicates a naval weapon – the recoil on board ship was checked by passing a rope through the cascable. Towards the breech end of the gun, at the top, is the vent, a drilling which passes down into the gun chamber.

Internally the barrel is bored out to the required diameter or calibre until the chamber is reached, at which point the bore is slightly reduced so that the projectile, when loaded and rammed down, stops in the same place every time so as to give a regular volume of chamber in which the gunpowder charge will ignite. Were the bore the same diameter all the way to the end, the projectile could be forced down on top of the powder in such a way as to possibly prevent ignition, or the volume would vary from shot to shot and thus give irregular performance.

To load the gun the charge would be prepared; in the early days this was simply a matter of scooping up whatever the gunner thought was the right amount of powder and shovelling it into the muzzle. It was then swept down into the chamber by a ramrod. In later years the gunner made up his powder into weighed flannel bags so that regularity from shot to shot was assured. This was placed in the muzzle and rammed home. Above this might go a wad, or the shot might be prepared with a wood "bottom" which acted as a wad; either way there would be some form of seal above the charge to ensure that when the charge was fired as much gas as possible was trapped behind the shot or shell and actually did some useful work in pushing the projectile out. A wad might also be rammed down on top of the projectile in the days of round shot in order to prevent it rolling back to the muzzle while the gun was being manoeuvered into position to fire. Finally, fine gunpowder from a powder-horn would be poured into the vent. The gun would then be laid and the gunner applied his match – a slow-burning cord impregnated with saltpetre – to the powder in the vent. He took good care to stand well away and apply the match at arm's length, for when the gun fired the recoil would drive the weapon back several feet.

After firing, the gunners would first manhandle the gun back to the firing point, and then apply a wet sponge rammer to the bore and scrub vigourously. This served two purposes; firstly it scrubbed away and removed the worst of the fouling always generated by the explosion of gunpowder, and secondly it was hoped that this would extinguish any smouldering residue of charge or bag or wad which might be lingering in the barrel, awaiting the loading of the fresh charge. Once the bore had been sponged, one gunner placed his thumb firmly over the vent, a process known as "Serving the Vent" and held it there while the gun was re-loaded. For in spite of the wet sponging, it was not unknown for a spark to remain in the chamber, and the ramming of the fresh charge might set up a current of air through the vent sufficient to fan the spark into life and ignite the fresh charge as it arrived, and generally impale the poor loader on his own ramrod when the explosion occurred. So to prevent this through-draught, the vent was served, and barrack-room tradition has it, the gunner with the rammer was allowed to lay it across the head of the vent-server if ever he neglected his vital duty.

16 **Typical Rifled Muzzle Loader** of its time, the carriage is of the utmost simplicity and the trail end is rounded to permit the whole equipment to slide freely over the ground on recoil.

The carriage which brought the gun to the scene of the action, in the case of field guns, was a simple two-wheel structure. The wheels would be joined by the axle-tree, and running back from the axle-tree was the trail, a simple pole-like structure to act as a third point of support. At the end of the trail was the "trail eye" a metal ring which allowed the gun to be hooked to the limber, a two-wheeled cart for carrying a small supply of ammunition. A handspike, a simple wooden shaft, would fit into the end of the trail and provide leverage for it to be shifted left or right as desired for pointing the gun. The front end of the trail swept up in two side-plates terminating in semi-circular supports into which the gun trunnions fitted, held therein by metal "capsquares" or plates. Between the breech end of the gun and the trail was the elevating screw, a simple screw passing through a hole in the trail and connected at the top to a lug on the gun. Revolving a threaded handle on the screw would lift it through the trail and thus thrust up the muzzle or allow it to drop. Finally, simple sights would be fitted; a pointer at the muzzle, or perhaps only a notch cut in the muzzle swell, and a post with a notch in the top at the breech end. This post would be held in a clamp and engraved with lines to indicate various ranges. By sighting across the top of the post to the front notch and the target beyond, the gunner gave directions to a companion manning the handspike to swing the gun left or right until the line was correct, and then, with the elevating screw would adjust the gun until sights and target were aligned. Any correction for wind would be applied mentally by the layer – there were still some aspects of gunnery which remained an art.

For guns of position – siege or coast defence weapons – the carriage would be rather simpler but generally more massive. For siege purposes mortars were often used. These were large-calibre short-barrelled weapons, firing always at high elevations in order to be able to fire from cover and pitch their projectiles across intervening walls to drop into the target area. For coast defence the weapons used were generally the same as those used by the navy of the day – four-wheeled truck carriages made up from two side-pieces carrying the trunnions and two transoms holding the sides together. The rear transom would take a quoin or wedge for adjusting the elevation, and a wheel on each corner completed the outfit. The wheels were not for manoeuvering, simply to permit the whole affair to recoil on firing. A heavy "breeching rope" would pass from the fortress walls around the cascable, and this was the control to the gun's rearward flight. After it had come to rest the gunners would step in with metal-tipped handspikes and heave the cannon after re-loading until it was once more poking out over the ramparts.

17 **Hotchkiss' System.** The brass cartridge case serves to seal the breech as well as to carry the ignition and the shell. The block merely supports the base of the case and holds the firing mechanism, operated by the trigger on the pistol grip.

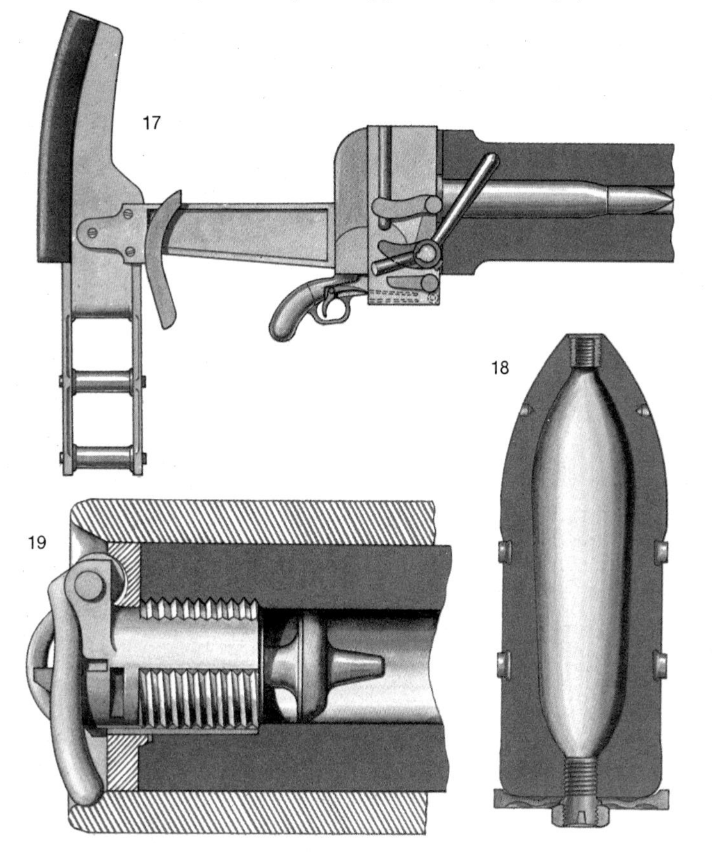

18 **Rifled Muzzle Loader Projectile.** In addition to its studs, for taking the rifling, a gas-check plate is fitted to the base. This corrugated plate will be flattened by the explosion of the charge and will be forced into the rifling grooves to seal the gas behind the shell and prevent erosion.

19 **The De Bange Obturation System.** This shows a very early application of the system, and the pad has been omitted for clarity. The vent bolt and breech screw can be seen, and the peculiar coned protrusion on the vent head was to position the cartridge correctly in the chamber so as to be ignited by the flash down the radial vent. It was found to be an unneccessary refinement and was soon abandoned.

Artillery remained at this pitch of mechanical development through the nineteenth century until the Crimean War, undisturbed by all the mechanical wonders of the Industrial Revolution. The steam engine had arrived, railways came into use, machines were brought into industry, but no trace of "fall-out" from any of this can be discerned in the design of ordnance. The Crimean War changed all this. It was the first war fought with the attendance of the electric telegraph and the newspaper reporter, and for the first time people in England who had never given a thought to their Army, or how it was operated or provisioned or armed found themselves in possession of full and scarifying details. While most people were exercised by tales of inefficiency and waste, blundering leadership and primitive conditions, one or two began to concern themselves with the apparent anomaly that the weapons of the Army had seen no change since the Peninsular War, in spite of the Victorian enthusiasm for mechanical improvements. It should not be thought that the fault lay entirely with the Army; they had been trying for some years to obtain a decent rifled breech-loading gun, but the only inventors who seemed interested were those whose knowledge of mechanical principles was outstripped by their enthusiasm. Although a design of breech-loader had been provisionally accepted during the Crimean War, it proved impossible to get a supply of guns made in time, and once the war was over the design was abandoned in the hope that a better one would shortly appear.

One of the men who decided that the present state of en-

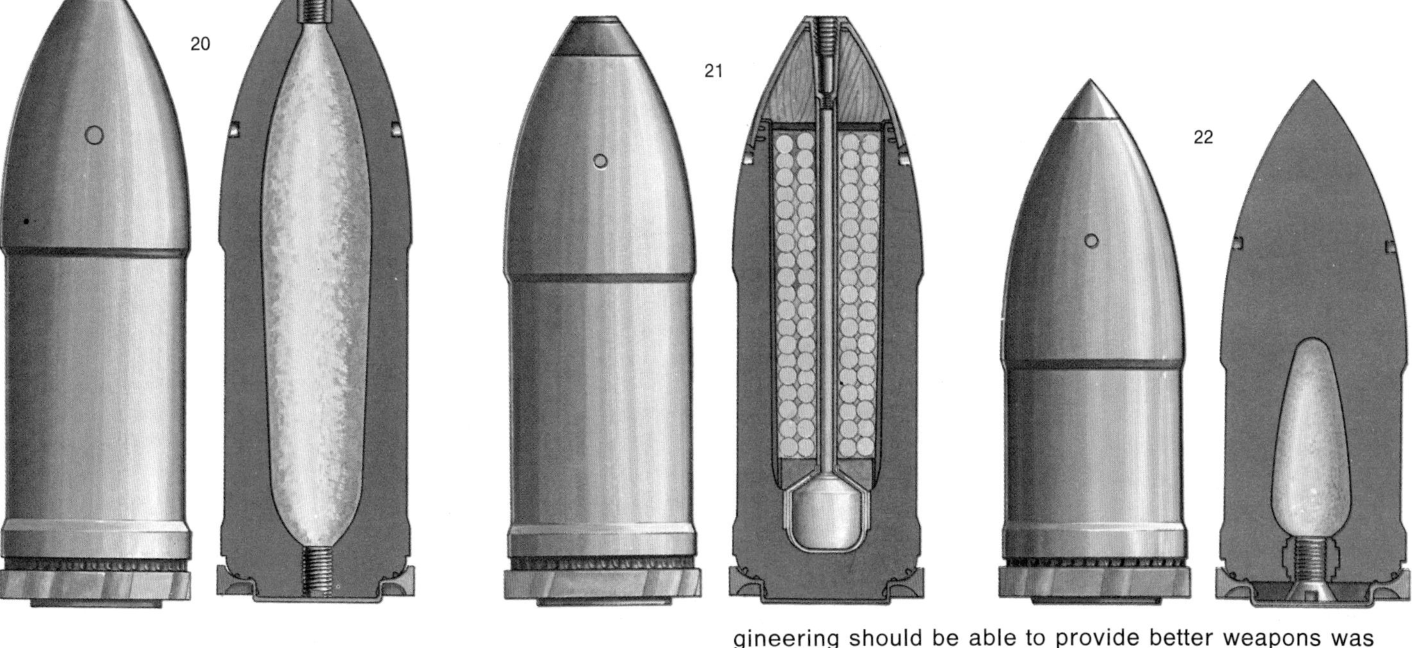

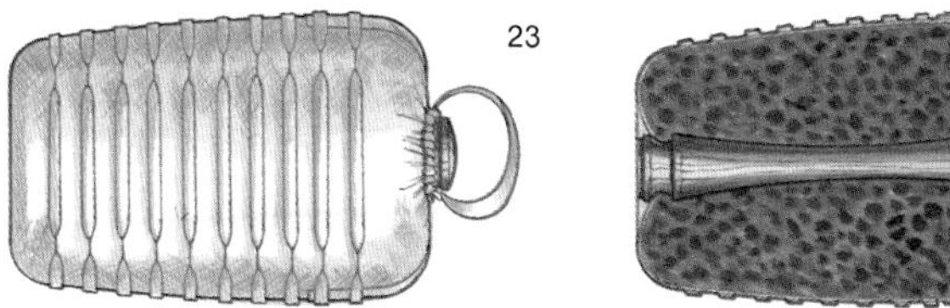

20 **Common Shell.** Filled with gunpowder and initiated by an impact fuze. Notice that this shell (as the previous two illustrations) is fitted with an expanding gas-check at the base, for use with rifled muzzle-loaders.

21 **Shrapnel.** Lieutenant Henry Shrapnel's invention took on a new form with the rifled gun. The head was pinned to the body, which carried a charge of lead musket balls packed in resin. A small bursting charge of gunpowder was located beneath the payload and was ignited by a time fuze flashing down the central tube. The resulting explosion drove the balls forward, forcing off the head.

22 **Palliser Shell.** Captain Palliser of the 18th Hussars invented this shell in 1864. By casting it nose down in a chilled iron mould, the head was hardened and an exceptionally successful piercing projectile was the result. Note the holes in the ogive for removal by the special tool.

23 **A Powder Cartridge.** For a rifled muzzle-loader, this example shows the central wooden former which kept the charge rigid and the tape loops around the bag which kept it in shape to facilitate loading.

gineering should be able to provide better weapons was William Armstrong, a Newcastle engineer who had made a name inventing and manufacturing hydraulic machinery. He had some new ideas on standardisation of manufacture and strength of materials, and he applied himself to the task of making a gun. He began by abandoning cast iron as a material of the past; wrought iron was the newer and more reliable substance, let us therefore make guns from it. Then, instead of piling on metal to give a factor of safety of umpteen, he carefully calculated the stresses likely to occur at various points on the barrel and designed it accordingly. Next he decided to improve the accuracy and rifled the bore with numerous grooves; in order that the projectile should be spun by the grooves he coated it with lead and made it longer in relation to its calibre. And since this system would entail difficulties in loading via the muzzle in the time-honoured way, he developed a system of breech-loading which was little removed from fifteenth-century ideas on the subject, but rather better engineered. (8). Eventually his gun was ready and after trying it out, he took it to London and offered it to the Master General of the Ordnance.

24 **Clerk's Recoil Buffer.** The first application of the hydraulic buffer was to the garrison slide and carriage for coast guns. As the carriage recoils, so the piston is driven into the cylinder of oil to absorb the energy. Notice also the traversing arrangements, powered by pulling on a rope around a capstan drum which moves the wheels by skew gearing.

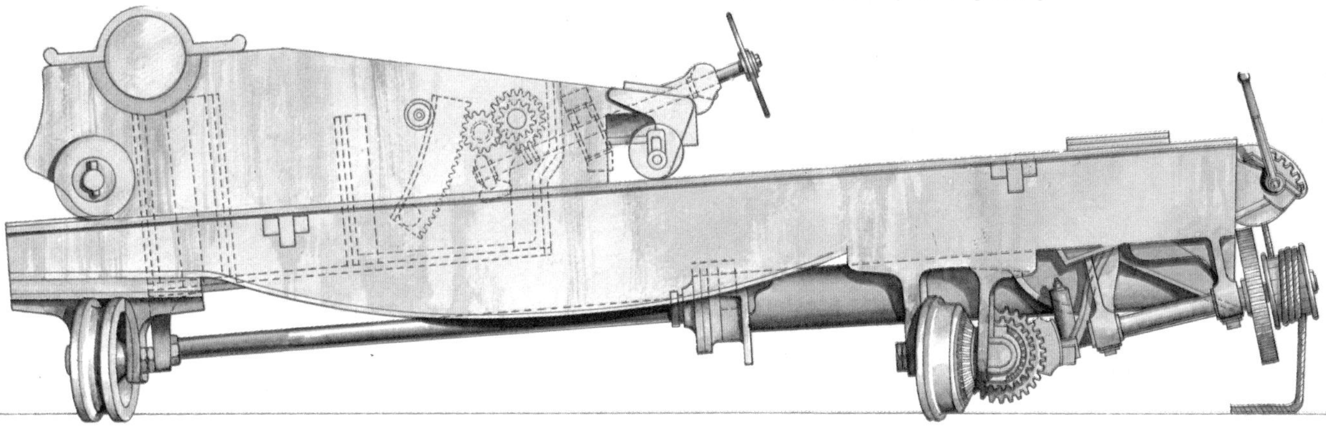

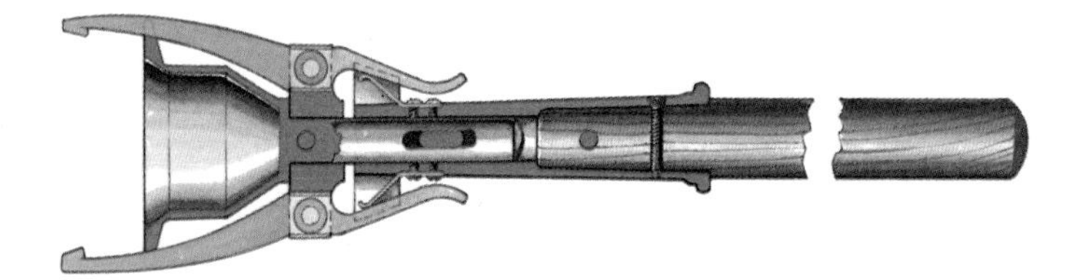

25 **Implement Removing Shell.** To unload a rifled muzzle loader it was necessary to reach down the bore with this device until its jaws clipped into the holes prepared in the shell nose. Pulling on the handle then tightened the grip and the shell could be drawn out.

His first gun, a 3-pounder, was delivered for trial in July 1855, and was favourably received. But in the way of all official bodies, it was felt that to accept one design without giving all the others a chance was unfair, and three years were spent in performing comparative tests of seven different types of rifled breech-loading guns. Finally, on November 16th 1858, the Special Committee, set up to conduct the trials, "had the honour to report and recommend the immediate introduction of guns rifled on Mr Armstrong's principle for special service in the field." Mr Armstrong, on hearing this, presented the Secretary of State for War with all the relevant patents as an outright gift. He was shortly thereafter appointed to the post of Superintendant of the Royal Gun Factory, Woolwich, where he could give his personal attention to the production of his designs. Within a year 162 guns had been produced, and by 31st March 1861 no less than 941 Armstrong guns had been produced for the Army.

The Navy had also accepted the Armstrong design, and by 1863 over two and half million pounds had been spent on new guns. But by this time a certain amount of complaint was being heard; the breech closing system tended to wear badly and become unsafe; the lead coating of shells stripped off in flight and the shells flew erratically; accidents occurred due to the breech piece being improperly secured. All in all, it was felt that the system was too delicate for service and too complicated for the simple soldiers to handle. Another point which had arisen was the arrival on the scene of the iron ship; the Armstrong guns, with their relatively weak breeches could not propel a shot hard enough to smash through iron plates, and there was a strong case for reverting to the muzzle-loader, at least until some better and stronger form of breech closure appeared on the scene.

Comparative trials were done once more, this time between muzzle-loaders and breech-loaders; but one thing had changed from the earlier muzzle-loaders – the advantages of elongated projectiles and rifling had been appreciated, and so the muzzle-loaders were rifled. Only three grooves were cut in the barrel, and those comparatively large in section; the shells were provided with two rows of soft metal studs, three in each row, so that as the shell was placed in the muzzle, the studs were engaged with the grooves, and as the shell was rammed home so the studs rode down the grooves (16). On firing the studs rode up the grooves and imparted the necessary spin to the shell.

The outcome of these trials was the selection of the Rifled Muzzle Loading (RML) system; the Rifled Breech Loader (RBL) and the RML were about equal as far as ease and rapidity of loading went, but the simplicity and strength of the RML system decided the issue, and the Armstrong gun's day was almost over. In all, 4540 Armstrong guns had been supplied, and they were retained in service for some years, gradually being replaced by the RML guns. Armstrong's other innovation had been in gun construction. Instead of taking one massive casting and boring a hole through it, Armstrong had used a number of thin tubes of wrought iron, shrinking them on to each other to provide the required strength and, by using varying lengths of tube, arranged for the barrel to be thicker at the chamber where the pressure was greater. This form of construction showed definite advantages in simplicity, weight and reliability over the crude casting, and the system was followed with the RML guns, with the addition of a solid forged breech plug. In order to make some use of old muzzle-loading smoothbores, a system of reaming out and lining them with a rifled liner was developed; the

26 **Loading Gear.** A Woolwich-built 12.5 inch muzzle loader on its Garrison Carriage, with a loading davit attached to the muzzle. It remained there during firing.

insert was a loose fit until expanded into place by the simple expedient of firing a heavy charge in the gun.

While all this was going on in England, similar moves were afoot in other countries. In the United States, Rodman and Parrott were applying science to gun design and laying down the basic principles of stress calculations, building their guns accordingly. Rodman placed every gunmaker in his debt when he developed a simple pressure gauge which could be inserted at varying points in the gun to demonstrate the development of gas pressure when the gun was fired.

In Germany, Krupp was working on the problem of producing gun barrels from solid blocks of steel, a proceeding which other gunmakers regarded with horror, for producing a flawless block of steel of such proportions was considered a minor miracle at that time. At the same time Krupp persevered with a design of breech closure in which the principal part was a simple block of steel sliding through a mortised hole in the breech end of the gun. The problem was to seal the escape of gas from the breech when the gun fired, both for safety's sake and for the sake of economy of effort – the more gas that leaked, the less was being used to push the shell.

In the following years the RMLs became bigger and more powerful, reaching their climax in the production of four 17.72 inch monsters, the barrels of which weighed 100 tons, provided for coast defence in Gibraltar and Malta. With a 460 lb charge these fired a 2000 lb projectile at almost 1700 feet per second, a performance which is far from negligible, and one which disproves the common belief that muzzle-loaders lacked power. But to achieve such performance the guns were getting longer and heavier and the point was being reached where complex mechanical contrivances were having to be introduced in order to load them, particularly on ships. Finally, in 1879, came the terrible accident on board HMS *Thunderer*. This ship was fitted with two 12 inch 38 ton RML guns in a turret. To load, the guns were trained inboard and depressed into two loading ports in the deck. Acting on orders transmitted by bells and indicators from the turret, the loading crew would ram home the 85 lb charge and 800 lb shell with a steam rammer. Withdrawing the rammers they would signal, "Loaded", and the guns would then train outboard, aim and fire in unison. On the fateful day, this drill was performed but one gun misfired. The noise and bustle in the turret prevented the crew from noticing the misfire, and as both guns were automatically retracted by hydraulic power ready for re-loading, the absence of recoil was also unmarked. The guns swung, dipped, and the telegraphs signalled "Load", which the loading crews did. The guns trained out, the order to fire was given, and the double-shotted gun blew apart, killing practically everyone in the turret and injuring many men who were standing on the deck watching the practice. The advocates of breech-loading pointed out that this sort of mishap was impossible with their system, which the advocates of muzzle-loading had, reluctantly, to admit. The *Thunderer* was the last straw; designs of breech-loader were once more demanded and this time the change was to be permanent.

The multitude of experimental designs of breech seal had by now calmed down to a few well-tried patterns. First came Krupp with his block sliding through a transverse slot in the gun body. Then came Nordenfelt who had produced a simple block sliding up and down in grooves in an open jaw at the end of the gun. He also was working on a rotating block which was to show itself in a famous context some years later. Armstrong's Elswick Ordnance Company had developed a heavy screwed plug which went into the back of the gun and carried a tin expanding cup on its front end; this sat in the chamber and was blown outwards against the chamber walls by the explosion of the cartridge. Finally a French inventor, De Bange, produced a screw breech with a resilient pad of asbestos, rape oil and suet squeezed between two metal surfaces so that the pad was forced outwards on to the chamber walls to form a gastight seal. Krupp and Nordenfelt demanded a metal case around the propelling charge to do the sealing, while their breechblocks merely formed a mechanical support; the other two systems were self-sealing and merely required the charge to be contained in a cloth bag.

Many and varied were the arguments which raged about these two basic systems – and still do for that matter. Suffice it to say that where speed is the prime requirement, then the metal case system wins hands down, while for heavy weapons the self-contained systems with bag charges are the most economical. By and large the dividing line falls at about six inches calibre; below that the cased charge is usual, above it the bagged type. But this is only a generalisation, and can be radically altered by national likes and dislikes or by the whim of the designer. In Germany, for example, bag charge guns were the exception, and cased charges have been used up to the very highest calibres.

By 1885, then, the breech-loading gun was back to stay;

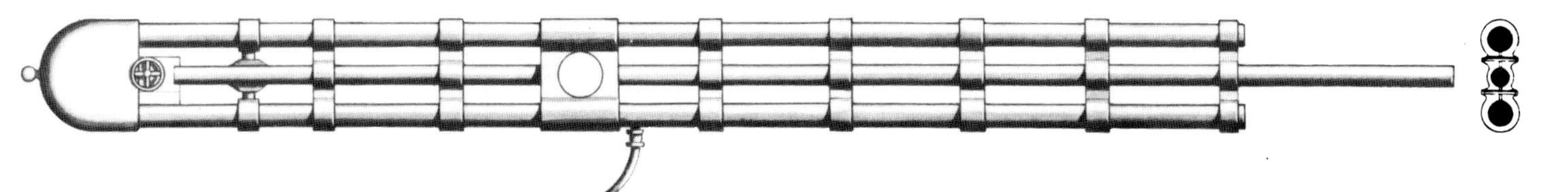

27 **The Zalinski Dynamite Gun.** This shows the plan view of the Zalinski gun as used on the USS "Vesuvius" and in the defences of New York Harbour at Sandy Hook. The two high-pressure air cylinders lie at each side of the long barrel, connected by fast-acting valves. When the shell had been loaded and the cylinders charged, the valves were thrown open to allow the air to propel the shell out of the barrel.

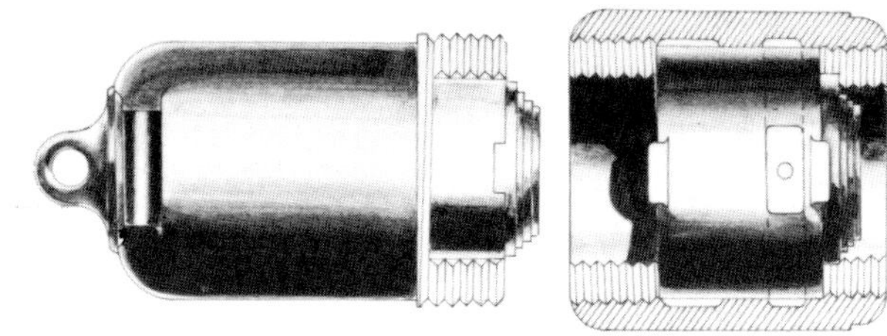

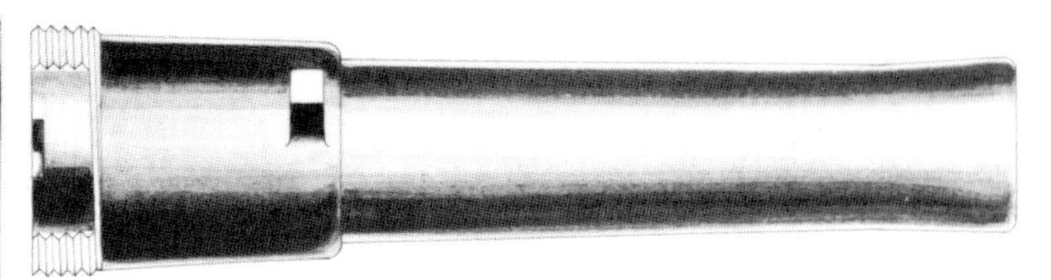

28 **A Jointed Gun.** One way to make a lightweight gun with a heavy punch is to dismantle it, a desirable feature when mule-pack artillery came into being. This version breaks into a breech and a muzzle section, with a third section which serves to hold it all together.

there were, of course, numbers of the old RML guns still in service in many countries, for they were quite efficient for many tasks and it would have been economically unrealistic to have scrapped them out of hand. But the breech-loader was the accepted norm, and in broad outline has remained much the same up to the present day.

While the problem of loading the gun from the rear end and the question of stabilising the shell in flight both seemed to have been solved, there were still one or two matters which invited attention from designers and engineers. Since the range of guns was steadily increasing, and with it their accuracy, some better method of aiming than the simple tangent sight was called for. At this time, remember, the gunner was still shooting direct – that is to say, he and the target were in sight of each other, and now that the target might well be four or five thousand yards away, hawk-like eyesight became a pre-requisite of a gunlayer. Another problem to be solved was the recoil of the weapon when it fired. Making guns more powerful meant that they recoiled more violently when they fired, and this meant that they bounded back several feet before coming to rest, and then had to be man handled back into the proper place before firing the next shot. There are numerous records of battles – Waterloo among them – where the gunners were so exhausted after a long fight that they could no longer heave the guns back, and fired from wherever they came to rest. If some method of absorbing this recoil could be devised, how much easier might the gunner's task become and how much greater his volume of aimed fire. The third question which deserved answer was the matter of protecting the gunners; modern small arms were beginning to develop greater range and velocity and killing power, and there was the danger that the gunner could be picked off by an infantry marksman before he ever got his gun into action.

The question of sights took its time; the direct fire concept was destined to stay in effect for many years to come, and therefore all that was wanted was a crutch for the layer's eyes. This was provided in the shape of Captain Percy Scott's Telescopic Sight, a simple telescope with cross-wires which was clamped into a suitable place on the gun carriage and coupled so that movement of the gun also moved the telescope. The basic idea was improved from time to time, but the original premise held good, and Scott's sight is virtually with us yet, in modern guise.

The business of absorbing recoil found many inventors willing to try, but only a few were on the right track. The Chevalier de Beaulieu approached the French Army early in the nineteenth century with proposals for placing a gas deflecting system on the muzzle of the gun so that some of the explosion could be diverted and used to push the gun forward to counteract the recoil force; he made it work with muskets, but on anything heavier it was less successful, and the muzzle brake had to wait for more favourable circumstances. Moreover this device, admirable and efficient as it has become, is no more than a recoil moderator – used by itself it was of no value, since the inertia of the gun carriage meant that the shot had gone and the gas dispersed before the carriage began to recoil, so that any muzzle braking effect had been dissi-

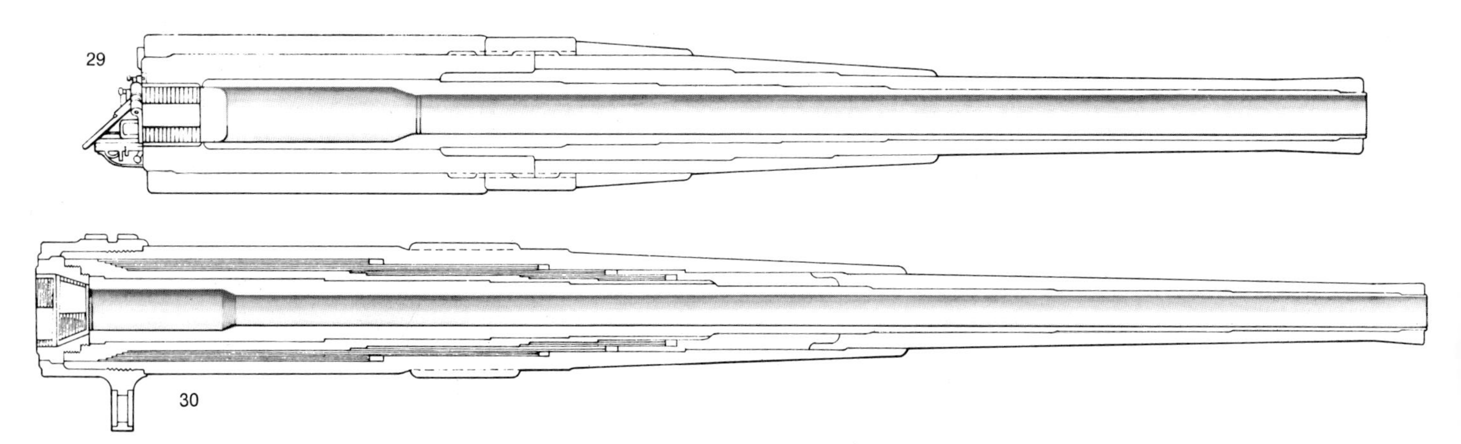

29 **A Built-Up Gun.** To counter the explosion force by scientific method rather than sheer bulk of metal, Armstrong developed this system of shrinking 'hoops' on to each other to place the whole structure in compression. The interlocking edges prevent the hoops from creeping longitudinally.

30 **A Wire-Wound Gun.** An example of the complexity of the gun-builder's art, this British gun shows the method of construction with interlocking hoops over an "A" tube, together with miles of wire tightly wound around the rear portion.

pated before the onset of the evil it was intended to counter.

In the late 1860s the British Government were suddenly made aware of deficiencies in their coast defences, both at home and throughout the Empire, and a vast programme of fort construction and armament began. The Royal Carriage Department at Woolwich were developing carriages for use in these forts, and one of the problems was that the space available in the casemates of the proposed forts was insufficient to permit a large gun to recoil on the current carriages. The carriage of the time was a simple inclined wooden platform up which the gun carriage proper – the four-wheeled ship carriage with small wheels – could slide. If the recoil seemed excessive, you threw sand on the slides; if it was insufficient and the carriage seemed liable to jump off the slide, you spread a little grease on it. In an endeavour to be a bit more precise about it, the Americans had developed the "Compressor" a collection of plates hung between the slide sides, interleaved with another sheaf hung from the gun carriage. By means of a screw-jack friction could be placed on the pack of interleaved plates so that when the gun recoiled the gun plates were dragged through the slide plates, and the amount of recoil could be moderately well controlled by the amount of pressure put on the pack. A quick-release was provided so that the friction could be instantly released to allow the gun to run back into battery after loading.

But the compressor was a primitive device, and one which demanded careful adjustment if it were to work properly; it would be much better if something could be devised to work without the necessity of the gunner having to judge how many turns of the screw-jack were needed for a particular shell-cartridge combination. The germ of the new idea came from outside the military establishment – as is often the case. The Superintendent of the Carriage Department at the time was at a scientific society soiree when he fell to discussing various matters with Professor Siemens, one of two German brothers who had set up business in London as engineers. Siemens suggested that the solution might lie in the application of hydraulics – then an up-and-coming branch of engineering, and the designers at Woolwich were put onto this the next morning. On investigation it proved that a hydraulic buffer had already been tried in the USA and was in use on some gun carriages, but it was used as the name implied, as a buffer. The gun recoiled freely up the slide until it struck the buffer piston rod; the impact drove the rod into the buffer cylinder where the piston head was forced against a body of water to absorb the recoil shock. All this did was to arrest the travel of the gun with somewhat less violence than before, but it did nothing to control or reduce recoil. Under the direction of Colonel Clerk of the Royal Engineers, the Woolwich designers developed a "buffer" (as they still called it, and do to this day) which was a step in the right direction. In this device a cylinder of oil was attached to the slide, and the end of the piston rod to the gun carriage (24). The piston head had a hole in it, and the recoil of the gun drove the piston through the oil at a speed controlled by the ability of the incompressable oil to pass through the hole to the other side of the piston. Suitable calculations, backed up by a certain amount of cut-and-try, determined the sizes of everything,

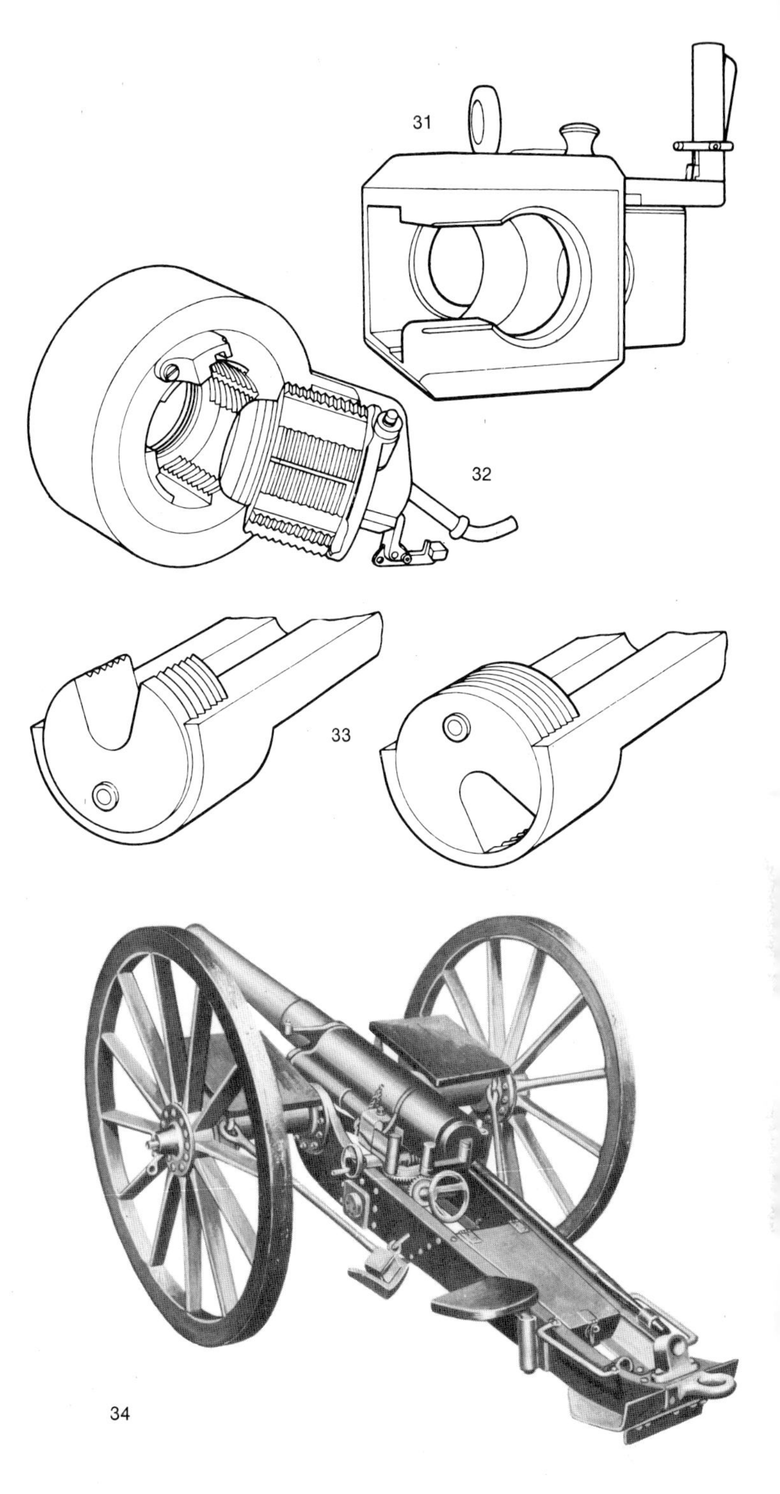

31 **A Block Breech Mechanism.** The breech of the US 75mm Howitzer, illustrating how the block slides aside to allow the round to be loaded. The firing mechanism is concealed within the block.

32 **A Screw Breech Mechanism.** The breech of the US 155mm Gun, showing the stepped thread screw and the obturating pad on the front end. The firing mechanism here is a simple hammer on the rear of the vent bolt.

33 **The Nordenfeldt Screw.** This system, used on the French 75mm M1897, uses a block mounted eccentrically to the bore's axis. The cutaway portion, when aligned with the bore, allows loading; rotating the block then swings a solid section into place behind the cartridge case.

34 **Maxim-Nordenfeldt 75mm Gun,** a commercial design sold to the Transvaal Government and used by the Boers during the South African War. With a hydrospring recoil system it was light and manœuvrable, but lively when fired.

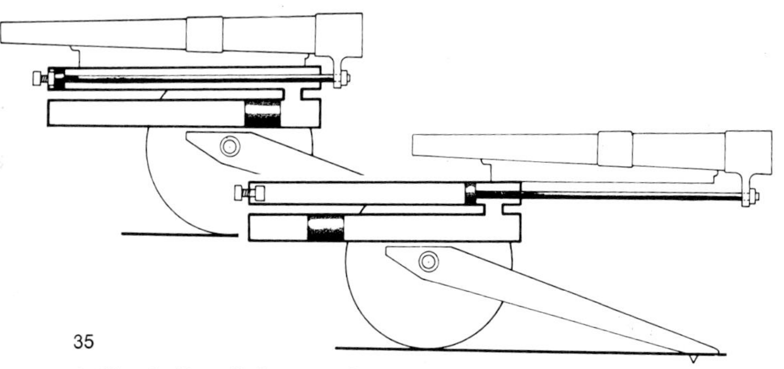

35

A Simple Recoil System. As the gun recoils so a piston is pulled to the rear, displacing oil into a second cylinder. The 'floating piston' is forced, by this oil flow, to move forward and thus compress air which is trapped in the space ahead of it. This cushions the recoil and also stores up energy to return the gun when recoil has ceased.

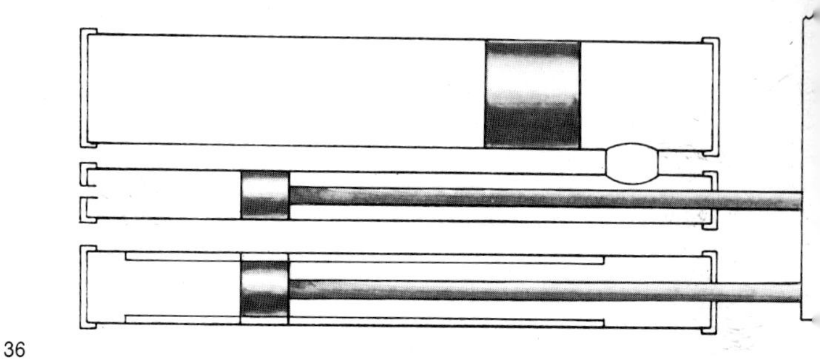

36

Independent Recoil System. Here the gun moves two pistons; the one in the lower cylinder acts purely to absorb recoil, while the one in the centre cylinder dislaces oil into the top, recuperator, cylinder to compress air via the floating piston. The buffer cylinder has a simple valve built into the piston head which gradually closes during the stroke to offer more resistance to recoil and finally to stop movement entirely.

and soon the first gun carriages with this device were built to be sent out to the newly-built Fort Cunningham in Bermuda. The virtue of this system was not only that it arrested recoil, but that it allowed the recoil to be controlled and delivered the recoil blow to the slide, and through that to the ground, in a diffused manner. This allowed the slides to be built less ponderously, making them easier to move, and also permitted the recoil to be controlled to a length which could be fitted into the casemates.

From this humble beginning, the hydraulic recoil system took off. In the original pattern, of course, there was no provision for automatically returning the gun to the firing position; it had to be manhandled back down the slide with handspikes, just as it had been in the sand and grease days.

From such developments, the concept of the Quick-Firing gun slowly took shape, and since this concept was to have appreciable effects on both the mechanics and the tactics of artillery for many years, it is worth examining in some detail. The object behind it was to produce a field gun which could be served quickly in action. To do this, certain features seemed desirable. Firstly the ammunition had to be in one piece – the cartridge enclosed in a metallic case carrying the ignition system, with the shell and fuze firmly attached, so that the gunner could load the gun in one movement. The breech mechanism had to be simple and quick acting. The gun, provided with some form of recoil control, had to stay in the same place for shot after shot, so that the gunners could remain in their places around it, ready to re-load and re-lay as soon as it had fired. And the gunners had to be provided with some form of protection against enemy small arms and shrapnel fire.

Many and varied were the attempts to solve this problem, and the honour of being the first successful designer goes to the French team responsible for the 75mm Model of the 1897 Field Gun. Although dated 1897 it was in gestation for many years as one design after another was tried and abandoned, but the designers' tenacity was well rewarded. The Model '97 became a legend in its own lifetime and many are still in use today, seventy years after its inception.

The "French 75", as it came to be universally known, incorporated firstly a hydro-pneumatic recoil system which not only absorbed the recoil but also returned the gun to the firing position after each shot. It had a quick-acting breech mechanism designed by Nordenfelt, used fixed ammunition, had a shield, and used a system of *"abatage"* in which the wheel brake shoes were dropped under the wheels when the gun was brought into action so that the carriage was held firmly in place. The simple pole-type trail had a spade at the end which, dug into the ground, also stabilised the carriage during firing. Finally it had an independent sighting system which allowed the gunner to continue laying the gun while an assistant layer on the other side looked after the matter of elevation. It was a *tour-de-force*, and it set the designers of the world on their ears when they heard about it. For a number of what would today be called "inspired leaks" from French Government sources made sure that the rest of the world were aware that the French had a world-beating gun, but at the same time strict security precautions also made sure that nobody knew exactly what were the various features of the weapon. It was carefully guarded when in transit and on manoeuvres no foreign observer or press representative was allowed near it. But for all that, snippets of information began to leak out, although most of them were exaggerated to the point where other nations were beginning to wonder whether the French really had a good gun or whether it was the Mitrailleuse all over again. The Mitrailleuse, a primitive form of machine gun, had been the secret weapon of the French in the 1860s, and had been as carefully guarded and as fearfully noised abroad. But in the Franco-Prussian War it had failed to live up to its promise, due largely to poor tactical handling, and, by and large, the Mitrailleuse set the machine gun back twenty years in the eyes of the world's soldiers. So there was an element of wait-and-see in response to the Model '97.

However, one thing seemed certain enough; the recoil system was the principal source of wonder. Observers reported on its uncanny steadiness in action, and the general feeling was that whatever else might or might not be the truth, certainly a successful recoil system had at last been developed. Admittedly, recoil systems which returned the gun to battery had already been seen, but they were generally combinations of a simple recoil buffer and a few large and coarse springs, which had the effect of choking off the recoil rather sharply and slamming the gun back violently in a manner which failed to absorb the recoil efficiently. The guns still leaped and rolled backwards, although not so far as before. But now they

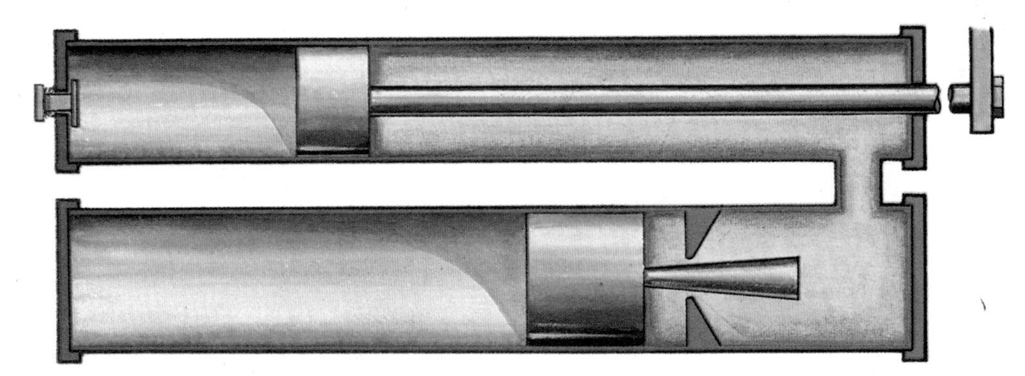

37

Dependent System. This system works on the same principles as the previous two but the floating piston is called on to do two jobs here; not only does it compress the recuperator air, but attached to it is a 'throttling rod' which, passing through a diaphragm, restricts the flow of oil and thus acts to absorb recoil and eventually stop movement. This was the system used on the French 75mm M1897 gun.

were saying that the gun recoiled across the carriage three or four feet before coming smoothly to a stop, then slid quietly and gently, albeit quickly, back into place, without the carriage moving an inch. It was all very mysterious and all very thwarting for gun and carriage designers.

The system adopted by the French was soon graced with the name "Long Recoil" and the world's gunmakers were hot in pursuit. Shortly after the gun was unveiled, Britain found herself involved in the South African War, and loud were the cries of both soldiers and men in the street at the absence of a good quick-firing field gun, until eventually Lord Haldane, Secretary of War, in a secret deal, purchased a number of 15-pounder guns from the German gunmaker Erhardt with which to equip a number of British batteries. This was by way of answering the critics, furnishing the Army with some modern guns so that they could formulate tactical doctrines, and, by the way, stimulating British gunmakers to try and produce something worthwhile, since they seemed to have become stuck in a rut. The US Army also purchased some Erhardt guns and, after modifying them slightly, went into production with their own variation which they christened the 3 inch Model 1903.

After the South African War, the Royal Artillery convened a number of committees to examine the reports of battles and, with these and verbatim reports of various artillery officers, came up with a specification for a new field gun and a new field howitzer, plus a smaller field gun for use by the Horse Artillery; since their role was to accompany the cavalry, they always required a somewhat lighter weapon. Gunmakers were circularised and eventually a number of designs were produced and tested. After critical examination the gun designed by Vickers was married to a carriage produced at Woolwich Arsenal, to produce the Field Artillery's 18-pounder (51) and a similar wedding produced the Horse Artillery's 13-pounder. In the howitzer field the result was a clear-cut win for private enterprise, the Coventry Ordnance Works' 4·5 inch howitzer being selected from a number of competing designs.

Germany too had taken to the drawing board after the French bombshell; Erhardt's 15-pounder employed a hydraulic buffer and spring return system but it was not taken up in any quantity by the German Army. Krupp now swallowed up Erhardt and produced a 77mm Gun Model 1906, which, as with the British designs, embodied all the now-standard features of the quick-firer. Other nations either designed their own or, more often, sought out the professional gunmakers such as Krupp, Vickers, Schneider, Cockerill or Bethlehem, and were supplied with an off-the-peg gun of the requisite type, or if they could afford it and were sufficiently fussy, a custom-built design which reflected their own idiosyncracies but in fact was probably little better than the stock model. Some tried to go it alone but had to call in the professionals before they were through, since gunmaking had become more scientific, and the days when a gun could be knocked up in the town square by half a dozen ironmasters and a master gunner were gone for ever. Now you had to have skilled designers, metallurgists, hydraulic engineers, explosive technicians and ballisticians; you needed a sizeable proof range, a team of research and development engineers, and some pretty smart salesmen; and above all you needed a massive engineering complex which could turn its hand rapidly from field guns to battleship guns to armour plate rolling to locomotives or what-have-you in order to keep working in what was, after all, a pretty seasonal trade. For once you had outfitted the

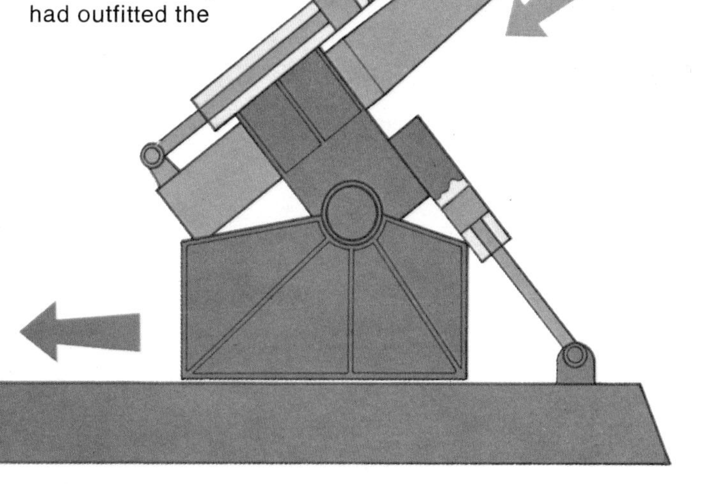

38 **Dual Recoil.** This system uses one recoil mechanism between gun and top carriage, and another between top carriage and basic structure. The representation here is purely diagrammatic.

Krasnian Army with 75mm field guns, you needn't expect them to be back with a return order for ten years or so, and the factories and the expensive help had to be given something to occupy their time. The answer was, of course, that you got your salesmen to work. You developed a better 75mm gun and got them to sell it to the Ruritanian Army. Then your salesmen went back to Krasnia and told them some spine-tingling stories about what the new gun could do, took the last year's guns in part-exchange, and re-fitted Krasnia with the new models, selling the old ones at a mark-down to one of the less belligerent Banana republics.

A good example of do-it-yourself was the Russian Army. At the turn of the century the arsenal at Putilov was turning out a home-designed gun which, for ingenuity, took some beating. The gun was slung in a top carriage which could slide back along the trail on recoil, and the recoil force was absorbed by a large number of rubber doughnuts which were compressed as the top carriage moved. The subsequent expansion of the doughnuts provided the power to return the gun to battery. Although it sounds comical in retrospect, and although no professional gunmaker would have been caught dead with such a design on his drawing board, the system worked reasonably well. The gun indeed was probably the best of its day, being more powerful and firing a heavier shell to a greater range than practically all of its contemporaries. But the recoil system demanded a long and ponderous trail, and the gun was not as stable as it might have been, so before the new century was very old a redesign to incorporate a more respectable recoil system was called for, and to achieve this the French firm of Schneider was asked to provide drawings. This they duly did, and also some weapons, but eventually, with a mixture of Schneider and Putilov features, a fresh carriage with hydro-pneumatic recoil control was put under the old gun which continued in service for many years.

By August 1914 the armament of the various world powers was more or less equal. Britain had the 13-pounder (3 inch) and 18-pounder (3.3 inch) field guns, Germany the 77mm, France the 75mm, Russia and the USA 3 inch models. All these fired shrapnel and had a maximum elevation of 16 to 18 degrees, and hence a fairly restricted range. One or two nations had also equipped themselves with field howitzers, augmenting the fire of their field guns with equipment firing a heavier shell capable of being dropped in plunging fire to search behind cover where the flat-trajectory guns could not reach. The French lost out in this respect; the Army had demanded a howitzer, well appreciating its value, but the purse-strings were held by the Chamber of Deputies, and one of their number, a M. Malandrin, rather fancied himself as a ballistician. He opined that howitzers were a waste of the taxpayer's money, and plunging fire could be obtained cheaply and easily by placing a steel washer over the shoulder of the shell so as to impede the airflow thus deforming the trajectory and making the shell fall more steeply.

There was a fair bit of tooth-sucking among the French gunners when they heard this, but, as is often the case, the paymasters had the last word and the gunners were given boxes of *"plaquettes"* and were told to shut up and get on with it. In the end, the pressure of war got them the howitzers they wanted, but the Malandrin disc soldiered on undisturbed until the Second World War came and the last of them were shot thankfully off.

The first surprise the Great War brought forth was the might of the German and Austrian heavy howitzers deployed against the Belgian and French frontier forts.

39 **Russian 12cm Howitzer.** Designed by Krupp, this weapon had been originally built in Essen in small numbers and sold to Russia in the hope of then selling them a licence to build their own. The outbreak of war removed the need for legal niceties and the howitzers were made at Putilov Arsenal.

The German 42cm howitzers (70) had been designed by Rausenberger of Krupps, beginning in the 1890s with a coast gun and gradually working his way, model by model, through the Greek alphabet, getting bigger as he moved on. His 42cm "Gamma" howitzer had to be carted piece by piece to its selected site and there assembled. The German Army felt this was a fine weapon; wherever they moved on the Continent they would come up against fortresses, and "Gamma" appeared to be the answer – but would Herr Krupp be kind enough to make it more portable? He did. And in the autumn of 1914 the Big Berthas rolled down the dusty roads of Belgium and pounded the Liege and Namur forts into rubble.

The Austrian "Schlanke Emma" howitzers (69) were called into assist the Berthas as the advance rolled on. These should have been no surprise to the rest of the world, since the Austrians hadn't been particularly security conscious, and details of these weapons had occasionally been seen in the Press. All the same, their appearance in battle caused a shudder to run through the Allies, and the prospect of producing comparable weapons was rapidly explored.

The British had been playing with the idea of a heavy howitzer for years. Experience in South Africa had inclined them to the belief that a few such weapons would be a useful reserve to deal with besieged towns and fortified places which the normal field weapons found too tough to crack. The Coventry Ordnance Works, who had produced the standard issue 4·5 inch howitzer now produced a 9·2 inch monster (75) which could be dismantled and moved piecemeal on wagon bodies towed by traction engines, and the prototype model was undergoing a leisurely evaluation at the Siege Artillery's range in Wales when war broke out. It was rapidly approved, packed up and sent off to France in time to make its debut at the Battle of Neuve Chapelle, while the Coventry factory were put to work turning out more.

Once this was under way, acting on the theory that if 9·2 inches was good, 15 inches couldn't fail to be better, Coventry scaled up their design to make a 15 inch howitzer, and offered this to the Army. But the Army were more worried about adequate supplies of field guns, and as far as they were concerned, the 9·2 inch Howitzer was quite sufficient for their needs in the heavy line. The managing director of Coventry Ordnance Works was Admiral Bacon, a retired naval officer, and he approached Winston Churchill, First Lord of the Admiralty, in the hope that Churchill might know some approach which might influence the Army. But Winston, in his own inimitable way, was more concerned with advancing the Navy's cause, and since there didn't seem to be much doing afloat, he was afire with the idea of putting sailors into the land battle, just to ginger up the soldiers. So he adopted the 15 inch howitzer, had some more made, manned them with Marines and sent them to France.

Eventually the Navy came to the conclusion that they were better at sailing than soldiering, and, once Churchill had left the Admiralty, set about tidying up all the odd units which were straying around Flanders. Admiral Bacon, who had in some devious fashion become a Colonel in the Marines and was commanding the 15 inch batteries, was recalled to seagoing duty and took over the famous Dover Patrol, and the howitzers were handed over to the Army. They were less than delighted, considering them a

41 **British 5 inch Howitzer.** This elderly warrior made its debut in the Nile Campaign of 1897, and numbers went to France in 1914. It was soon replaced by the more efficient 4.5 inch Howitzer and was then used for training, though a number were sent to Russia in 1916.

40 **British 6 inch Mark 7 Field Gun.** During the South African War a few naval 6 inch guns were mounted on locally-made carriages and sent to the front. After the war some properly designed carriages were provided but these had too little elevation to be useful, and a fresh design, shown here, was developed to give better range. Over a hundred were built, but at 25 tons weight they were far too heavy and were eventually replaced by a more modern weapon which weighed fifteen tons less.

waste of money, firing a poor shell to a range which was hardly commensurate with the amount of work demanded in emplacing the weapons. As soon as the war was over they were scrapped with almost indecent haste.

Two other types of weapon came into their own during the First World War, the railroad gun and the anti-aircraft gun, and we will look more closely at these in later sections devoted to more specialised weapons. In the broad field of artillery in general, the major advances during the war years were in such recondite areas as fire control, ballistics, specialised ammunition, calibration, and more versatile carriages. In the first place the nature of the war had driven the gunners out of sight. No longer did they sit in the open and engage the enemy in plain view. The guns were concealed behind hills, in woods, in sunken roads, in pits, anywhere where they were invisible to the enemy. This meant that the observer at the front was the only man who saw the target and actually guided the gunner's aim. To the outsider it always seems a minor branch of one of the Black Arts that the gunner appears to blast off into nowhere at an invisible target, and yet the shells drop where they should. It's all very easy when you know how it's done, and the secret lies in the Panoramic or Dial Sight. This can be compared to a periscope, the head of which can be revolved in a complete circle, and this head is given a scale to indicate the relationship between the line of sight through the head and the axis of the gun's barrel. The sight is clamped to the gun carriage with its eyepiece parallel with the barrel; when the periscope head points forward, also parallel

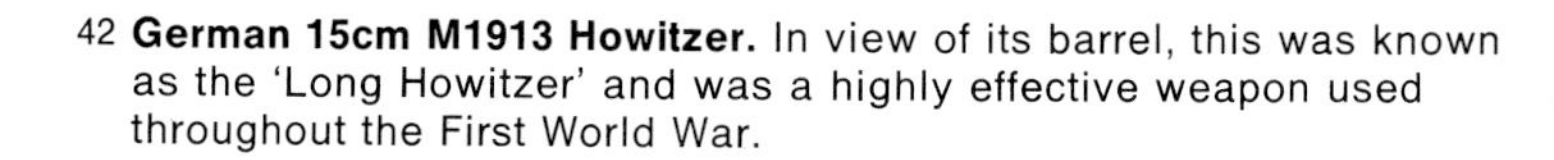

42 **German 15cm M1913 Howitzer.** In view of its barrel, this was known as the 'Long Howitzer' and was a highly effective weapon used throughout the First World War.

43 **Austrian 15cm M1914 Howitzer.** This was introduced to replace the Model 1899. Several hundred pounds lighter and much easier to handle, it outranged the earlier model by almost two thousand yards.

with the barrel, the scale reads 0. With the periscope revolved and pointing back over the eyepiece, the scale reads 180 degrees (or 3200 mils, or 200 grades or 14 quarts 1 pint, or whatever you care to graduate it in).

Now we take a map and mark on it the target and the gun and draw a line between the two. This we label 0 degrees, irrespective of what its real bearing happens to be. Next we find a prominent object visible from the gun and marked on the map. Join gun and object on the map with another line and measure the angle between this line and the original "zero" line. Call the object the "Aiming Point" and call the gunners together and point it out to them. Then chase them back to their guns, tell them the angle you discovered from the map, and let them swing the periscope head of their sight so that it points away from the barrel axis by the necessary amount. Now they heave and traverse the guns until the sight is looking at the "Aiming Point", and the barrels will be pointing along the line to the target.

But what if there isn't a prominent object, did you say? Well, never mind. There are several other ways of doing it, as long as you have a compass handy. I won't go into them, or this will turn into a Drill Manual; besides, we have to have a little bit of mystery left, or people will be going round thinking that gunnery is easy.

Calibration is a high-powered word which means taking into consideration the fact that the gun wears out a little every time you fire it, and therefore it gradually loses velocity and range. To calibrate a gun we fire it at a known range and, by correctional adjustments to the sights, fix it so that when, say, 8000 yards is set on the sight, the shell actually lands 8000 yards away. The muzzle velocity can be measured and guns grouped so that all the guns of a troop or battery are firing at the same velocity and range, and thus the shells fall together and make a more effective crump at the target without the necessity for complicated individual corrections.

In the ballistic field, the war, with its demands for accuracy due to shooting close to the front-line troops, demonstrated the need to take into account the effect on the shell's flight of such things as wind speed and direction, humidity, barometric pressure and temperature. An artillery meteorological service came into being which supplied the necessary data at regular intervals. With the results of this type of calculation it became possible to "predict" fire, in other words to calculate the range to a target, apply all the corrections and then open fire and hit the target straight away without having to fire ranging shots in order to eliminate the ballistic effects. This improved the surprise element and made lengthy preparatory fire, which inevitably warned the enemy that something was afoot, no longer necessary.

In the ammunition field, much development took place in order to try and tailor the shells to the tasks demanded of the gunners. The shrapnel shell, long considered the standard field artillery projectile, was a fine weapon for use against troops in the open, but once trenches and fixed defences concealed the targets it was found wanting. The prospect of putting high explosives into shells had tantalised inventors ever since high explosives had existed, but the difficulty to be overcome was the sudden and enormous acceleration as the shell was fired. The early explosives, nitro-glycerine, dynamite, blasting gelatine, were all of far too nervous a disposition to put up with such treatment and would immediately detonate and destroy the gun.

An early attempt to circumvent the problem was the use of compressed air as the propelling medium. Originally proposed by a Mr Mefford of Chicago, he employed a Lieutenant Zalinski of the US Navy as his Public Relations man to such good effect that within a short time it became the Zalinski Dynamite Gun. The "dynamite" side of it came from the filling used in the shell, but the propulsion was entirely by compressed air, with two large cylinders alongside the barrel containing 2000 lbs per square inch of air (27). The finned shell was loaded into the barrel and then valves were thrown open to admit the air behind

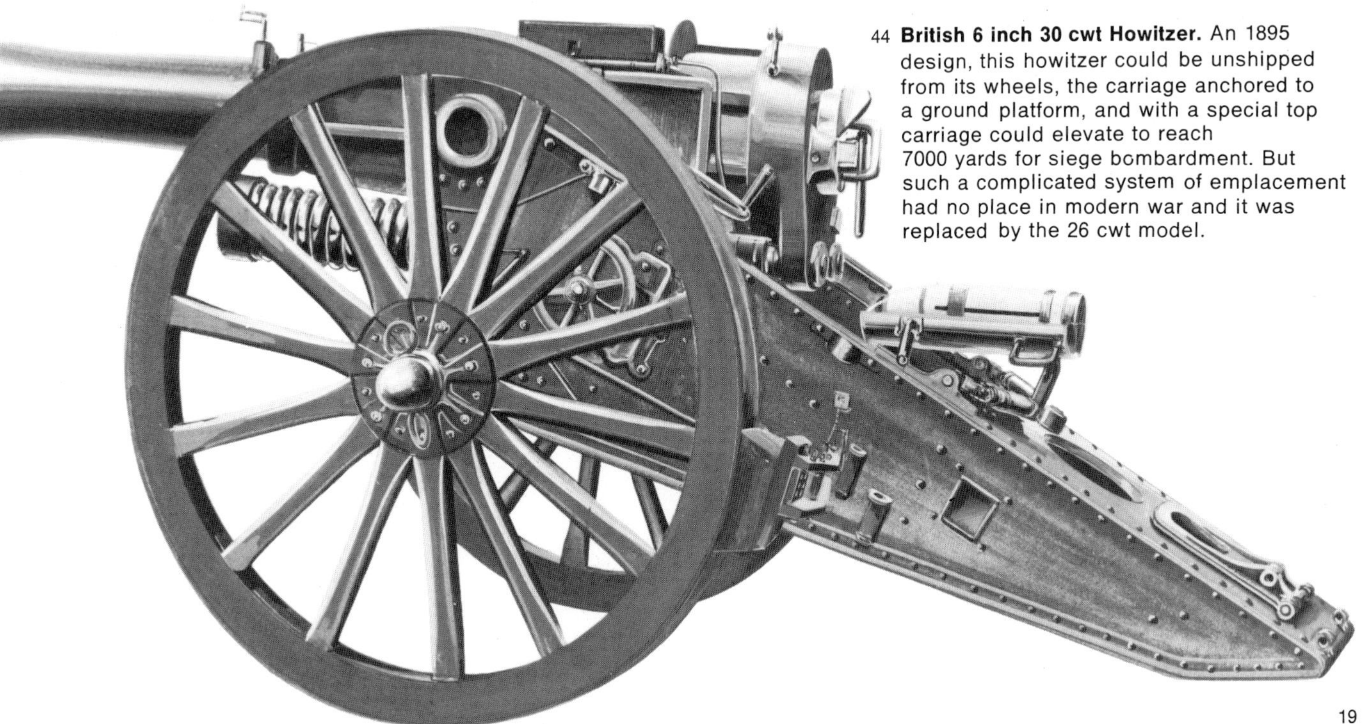

44 **British 6 inch 30 cwt Howitzer.** An 1895 design, this howitzer could be unshipped from its wheels, the carriage anchored to a ground platform, and with a special top carriage could elevate to reach 7000 yards for siege bombardment. But such a complicated system of emplacement had no place in modern war and it was replaced by the 26 cwt model.

45 **Austrian 104mm Field Gun.** A powerful weapon which could outrange most of its contemporaries, it was over-weight and the long trail made it an awkward item to manoeuver.

46 **British 6 inch 26 cwt Howitzer.** Developed in 1915 to replace earlier 6 inch howitzers which were too heavy in the Flanders mud and which had insufficient range, this weapon remained in service until the middle of the Second World War.

47 **British 2.75 inch Mountain Gun.** An improved model of the original 10-pounder "Screw Gun", the 2.75 inch came apart into six mule loads and was the basic weapon of the Indian Mountain Artillery. Introduced in 1914 it had a short life, being superseded by the 3.7 inch Mountain Howitzer as soon as the war was over.

48 **German 7.7cm Field Gun M1916.** Replacing the models of 1896 and 1906 this had a longer barrel, higher velocity and longer range, together with the ability to reach a higher elevation, but the cost of it all was a twelve per cent increase in weight.

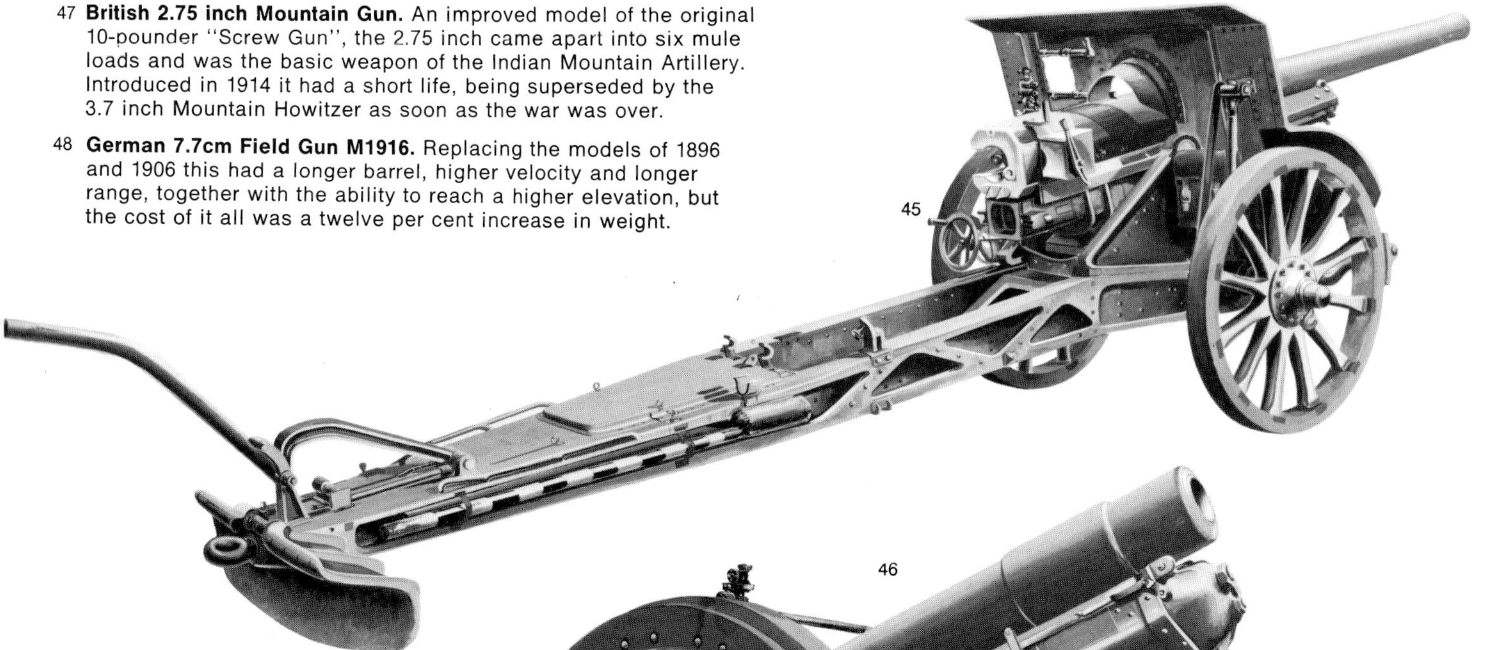

45

46

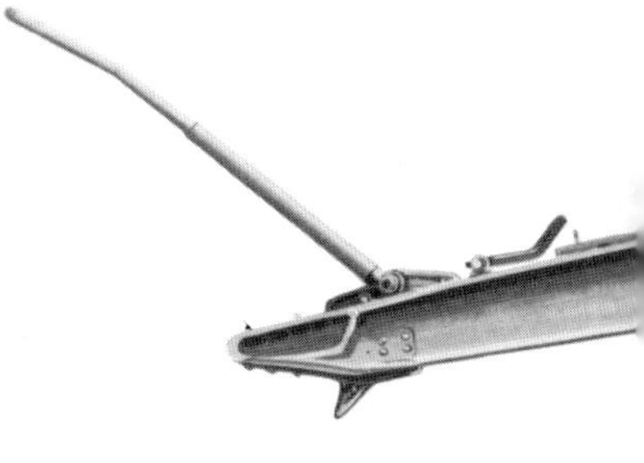

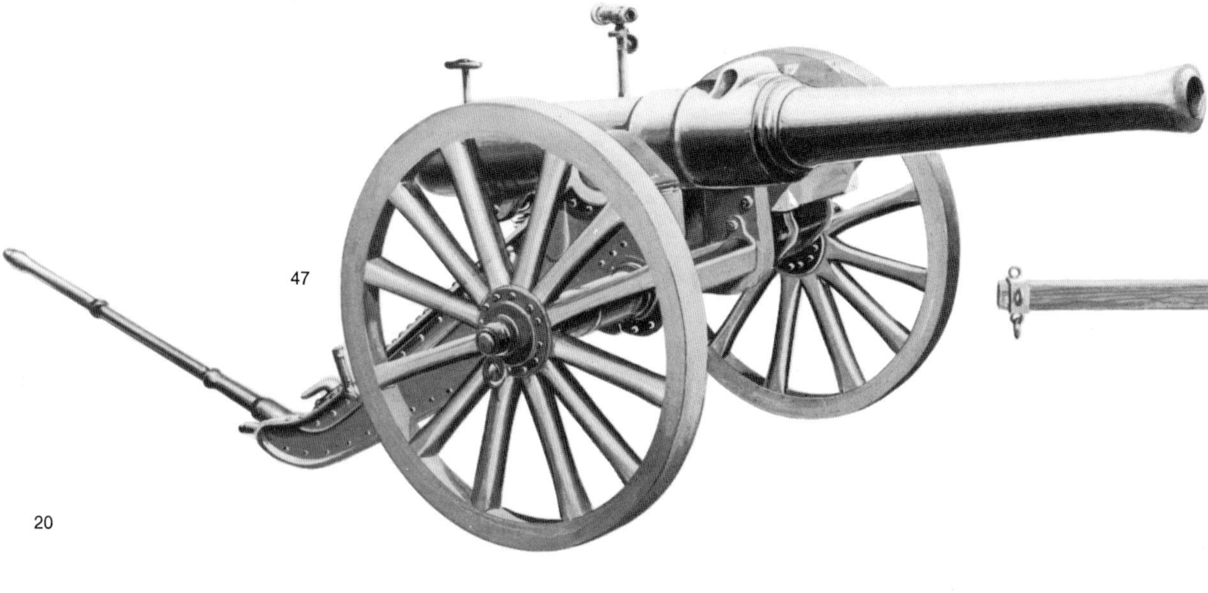

47

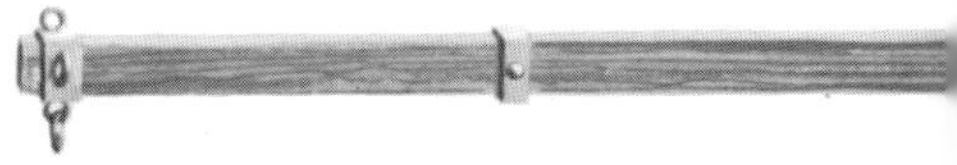

it, thus launching the shell with a comparatively gentle acceleration.

Improbable as it all sounds, the Dynamite Gun was taken up; the US Army installed some at Sandy Hook to defend the entrance to New York Harbour and the USS *Vesuvius* was equipped with a sea going version. In 1890 one was purchased by the British Army and installed at Dale Fort on Milford Haven. An old iron paddle steamer, the *Harpy*, was obtained to serve as a floating target, and trial firings took place in 1893. The gun was 15 inch calibre and fired a 10 inch subcalibre shell of 493 lbs to 4000 yards. The explosive shell functioned satisfactorily, to the detriment of the *Harpy*, but the eventual conclusion was that the amount of machinery and effort involved was out of proportion to the results, and the gun was dismantled and scrapped.

In the US they survived a little longer, the *Vesuvius* actually firing nine rounds in Manila Bay during the Spanish-American War, but by then the conventional gun could outperform the Zalinski Gun, and, moreover, the high-explosive problem was on the way to being solved. By 1900 the Dynamite Gun was a thing of the past.

The explosive which *would* stand up to being fired from a gun was Picric Acid. This was a chemical which had been used in the dye industry for years, and, as might be assumed, was quite stable. But when suitably roused by a detonator it could be made to detonate violently. Many secret experiments were made by the British Army at the Siege Artillery range at Lydd, in Kent, and the eventual solution was to melt the acid crystals and pour them into the shell. Then a small bag of crystals was placed alongside the fuze detonator. On the fuze being initiated, the more sensitive crystals detonated and they in turn would set off the main charge. To conceal the precise nature of the explosive, and also to commemorate its birthplace, it was christened "Lyddite". But security in those days was a little lax by modern standards, and within a short time almost every nation was using the same substance under a different name – Melinite, Shimose, Granatfullung 88 – it was all picric acid.

Lyddite was a start. It was not entirely satisfactory, being prone to incomplete detonation, and Germany began to look for something better, followed soon by Britain. Their solution was TNT, but it took a lot of experimenting to develop a suitable system of initiation to ensure thorough detonation. By 1914 TNT shells were beginning to form part of the issue for howitzers and a very few were available for field guns, but under the conditions of the Western Front the proportion began to increase until by the end of the war the HE shell issue outnumbered the shrapnel by a considerable amount.

When the Duke of Wellington wanted to conceal troop movements he had damp straw burned upwind. In 1914 something quicker and more reliable was needed, and the smoke shell came into being. Filled with white phosphorus, this burst on impact to release the chemical filling which ignited in contact with the air and generated a thick cloud of white smoke. Later this basic design was revamped to take poison gas and liberate that in a similar

48

49

49 **Austrian 15cm M1899 Howitzer.** This medium field howitzer ranged to almost 7000 yards but when ready for transport, as shown here, was a cumbersome load for a horse team.

way with more lethal effect.

After the smoke cleared away in 1918 things were never to be quite the same in many respects, and the artillery were among those who felt the wind of change blowing round their ankles. The big gunmakers – Vickers, Cockerill, Krupp, Skoda, Bethlehem and their ilk – were no longer faced with a seller's market. When the vast European armies were dismantled, enough artillery equipment to equip every standing army in the world flowed onto the market and the gunmakers were obliged to turn their hands to other products. Also military design establishments had been increased during the war and though they were now reduced their final size was still well above the pre-war level and they had tasted the sweet fruit of producing their own weapons. Seeing that they could do as well as or better than the professionals they busied themselves with designs while the gunmakers made locomotives and office chairs and razor blades.

The Americans were particularly active at this time, producing many designs, particularly of anti-aircraft guns and self-propelled guns, which formed a useful foundation for future work. Most of these projects were shelved for lack of money, a situation which faced every army at the time, and many failed to survive their stay in cold storage. But a number of the weapons deployed by the US Army during the Second World War could trace their parentage back to the experimental efforts of 1921.

Another result of the war was that almost every nation convened some sort of post-mortem Board to consider the lessons that had been learned and to prognosticate the size and shape of future artillery. In the USA the Westervelt Board sat in 1919 and made far-reaching recommendations which laid down an artillery equipment policy which has been followed almost to the present day. In Britain it was less formally organized, but by the middle 1920s some figures were being put on paper to point the way to the next generation of field guns. By 1930 a 4.1 inch howitzer had been developed by Vickers to meet a specification for a field artillery equipment, and a handful were made for trial. But the financial wizards were somewhat dismayed to hear of the possibility of adopting an entirely new design. What was to happen to the hundreds of perfectly serviceable 18-pounders the Army already owned? Unthinkable to scrap them out of hand. And under this sort of pressure the ideal requirements were watered down until the final answer was a gun of a size which would allow its barrel to be slipped into the jacket of the 18-pounder equipments to save the cost of new carriages. And that is why the famous 25-pounder came to be 3.45 inches calibre.

Nothing daunted, the designers went on to develop a new carriage which would allow the maximum performance to be extracted from the gun, since the 18-pounder carriage was not strong enough to stand the full charge cartridge designed for the 25-pounder. Some outstanding designs were mooted; one, which reached the mock-up stage, was a four-legged, all-round-traverse design which was almost the same as the Skoda FH43 design of 1943. But the money question was all-important. A simple split-trail carriage was decided upon. This was to be sent to the School of Artillery for trial by the customers, and in order to give the testers something to use for comparison, a few guns were mounted on the old 4.1 inch howitzer carriage. A feature of this carriage was a round platform which, slung beneath the trail, could be dropped to the ground and the gun pulled on to it so that the gun wheels ran round the platform's rim and allowed 360 degree traverse; this was an improved version of Capt. Hogg's original gun-wheel platform, developed as an anti-tank measure in 1918 (160).

To everyone's astonishment, the testers rejected the split-trail and unanimously decided that the 4.1 inch carriage with its platform was the one they wanted. By this time (it was now 1936) the writing was on the wall, and instead of a long wrangle in the hope of changing people's minds, the recommendation was accepted and the 25-pounder carriage design was fixed. In the event, the Second World War arrived before the carriages, and the 1940 battles in France were fought with the 25-pounder Mark I on the 18-pounder carriage.

In Germany the designers had a less difficult task. The Versailles Treaty with its limitations on military strength and armament had ensured that the Army was not burdened with a vast stock of wartime weapons which had to be used up to keep the paymasters happy, and when the expansion of the Wehrmacht began, new designs of gun were immediately called for. Here there were no Government design agencies, as in other countries. The Army relied on the gunsmiths, who had kept their skills alive during the lean years by forming trade alliances with foreign firms, or setting up "front" companies in other countries. Their designers went abroad and kept busy until they could be recalled to Germany and put to work, and in their years of exile they drew, tore up, redrew and polished a number of designs which were then ready to be put into production at short notice. They had also explored various new theories in gun and carriage design and laid the groundwork for several new and startling weapons which were to be revealed in the subsequent years.

The US Army in the lean years was making do with its World War One weapons, modified to allow high-speed towing behind trucks. In most cases this meant no more than fitting wheels with pneumatic tyres, but an alarming device was proposed for the 75mm M1897 gun; a sort of small-wheeled trolley into which the iron-shod gun wheels were run and locked for high-speed towing. But towards the end of the 1930s the designs were settled and the rising menaces of a re-arming Germany and a resurgent Japan effected a loosening of the purse-strings. The 105mm howitzer which had been recommended in 1918 was finally approved as were the 155mm gun and 155mm howitzer all of which were intended to replace the ex-French designs left over from 1918.

The eventual field guns – by which is meant the basic divisional weapon – of each of the major powers in 1939 was roughly comparable. The British 3.45 inch 25-pounder fired its 25 lb shell to 13,400 yards when fitted to its proper carriage (but only 12,800 when on the 18-pounder carriage), and weighed 1.8 tons. It was the smallest in terms of calibre and shell weight, but its range and manoeuverability were sufficiently vital factors to excuse this. The USA and Germany both used 105mm howitzers as their standard weapon. The US M1 fired a 32 lb shell to 12,200 yards and weighed 2.25 tons, while the German

50 **French 75mm M1897 Gun.** The gun which introduced the QF concept to a startled world. The Nordenfeldt breech screw is well shown here, as is the box trail construction which restricted the gun's elevation.

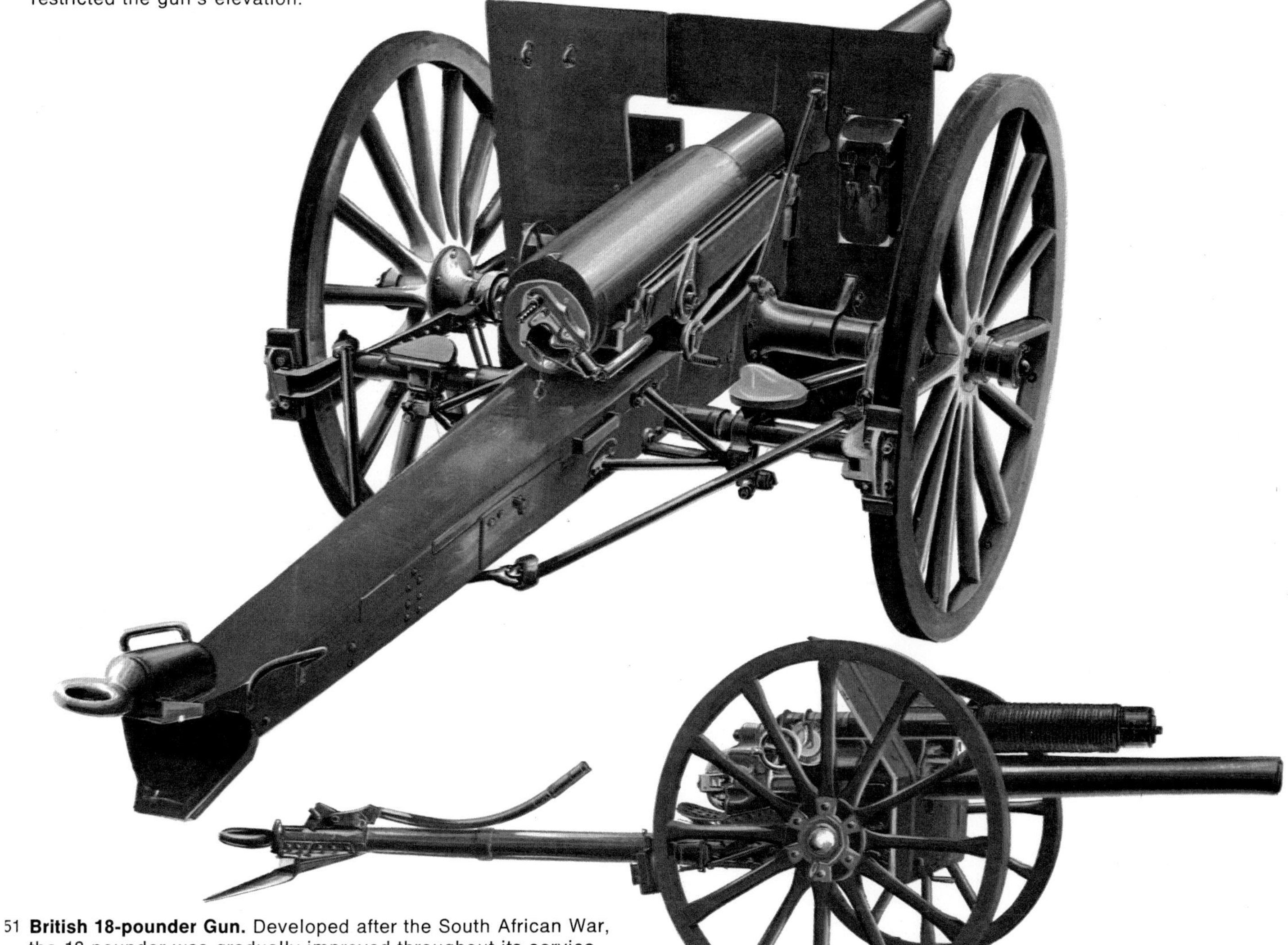

51 **British 18-pounder Gun.** Developed after the South African War, the 18 pounder was gradually improved throughout its service. It was the backbone of the British field artillery in France during the First World War and, fitted with the 25-pounder barrel, returned there in 1939.

leFH18 fired a 33 lb shell to 10,300 yards and tipped the scales at 1.9 tons. All these guns used propelling charges which could be adjusted so that they could function as high-velocity guns or low-velocity howitzers at will, and each had a selection of ammunition types available – HE, smoke, anti-tank, illuminating and so forth, sufficient to answer any tactical demand. The French still had vast numbers of 75mm M1897s, slightly modernised but by this time hopelessly out of date. The Soviets had a useful 76mm gun which ranged to no less than 15,260 yards and weighed only 1.6 tons, but its shell weighed only 14 lbs and could not compare with the Western guns in terms of power or lethality.

As the Second World War unfolded, the soldiers demanded more performance, more range, more lethality of their guns and shells, and the engineers, ballisticians and designers did their best to provide them. Before the war ended the combatant nations had, between them, developed almost four hundred different guns and howitzers ranging from 28mm to 800mm in calibre and on every conceivable type of mounting. So far as conventional guns went, the line of development was what one might expect: improvements in ammunition in order to extract the utmost performance from existing designs; then a new design intended to develop whatever tactical feature was uppermost at the time – shell weight, range, mobility, flexibility – the order of precedence varied with the various phases of the war – more ammunition development, then a new design, and so the pattern went on. The Western Allies were least affected by the pattern; they seemed happy to see the war out with the same basic weapons with which they had begun, although the British 25-pounder was given a muzzle brake to reduce the strain on the carriage when firing anti-tank shot, and was later equipped with a re-designed carriage with a narrow axle and hinged trail. This, a marriage of Indian and Canadian ideas, enabled it to be loaded into the ubiquitous Dakota transport aircraft without dismantling, and by breaking the trail in the middle, added another 30 degrees to the maximum elevation obtainable.

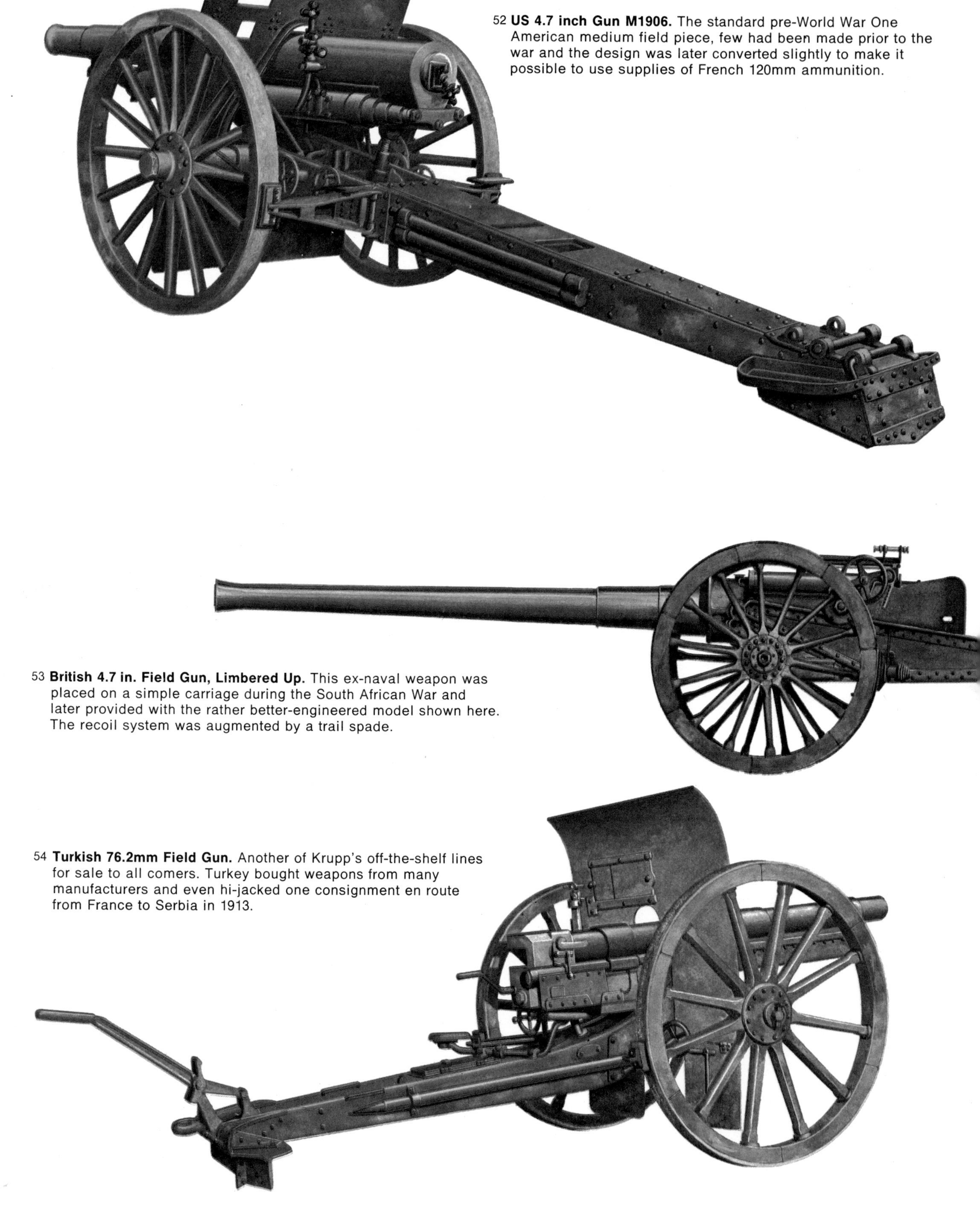

52 **US 4.7 inch Gun M1906.** The standard pre-World War One American medium field piece, few had been made prior to the war and the design was later converted slightly to make it possible to use supplies of French 120mm ammunition.

53 **British 4.7 in. Field Gun, Limbered Up.** This ex-naval weapon was placed on a simple carriage during the South African War and later provided with the rather better-engineered model shown here. The recoil system was augmented by a trail spade.

54 **Turkish 76.2mm Field Gun.** Another of Krupp's off-the-shelf lines for sale to all comers. Turkey bought weapons from many manufacturers and even hi-jacked one consignment en route from France to Serbia in 1913.

55 **German 21cm Howitzer.** Typical of the German heavy artillery of 1914-18 this 21cm weapon saw extensive use on the Western Front.

56 **British 60-pounder Gun.** This 5 inch calibre gun was designed in an attempt to produce the most powerful weapon possible within the weight limitations of what a horse team could pull. Although a success, the transport conditions of the First World War were more formidable than the designers had ever contemplated, and the prime mover was usually a Holt tractor.

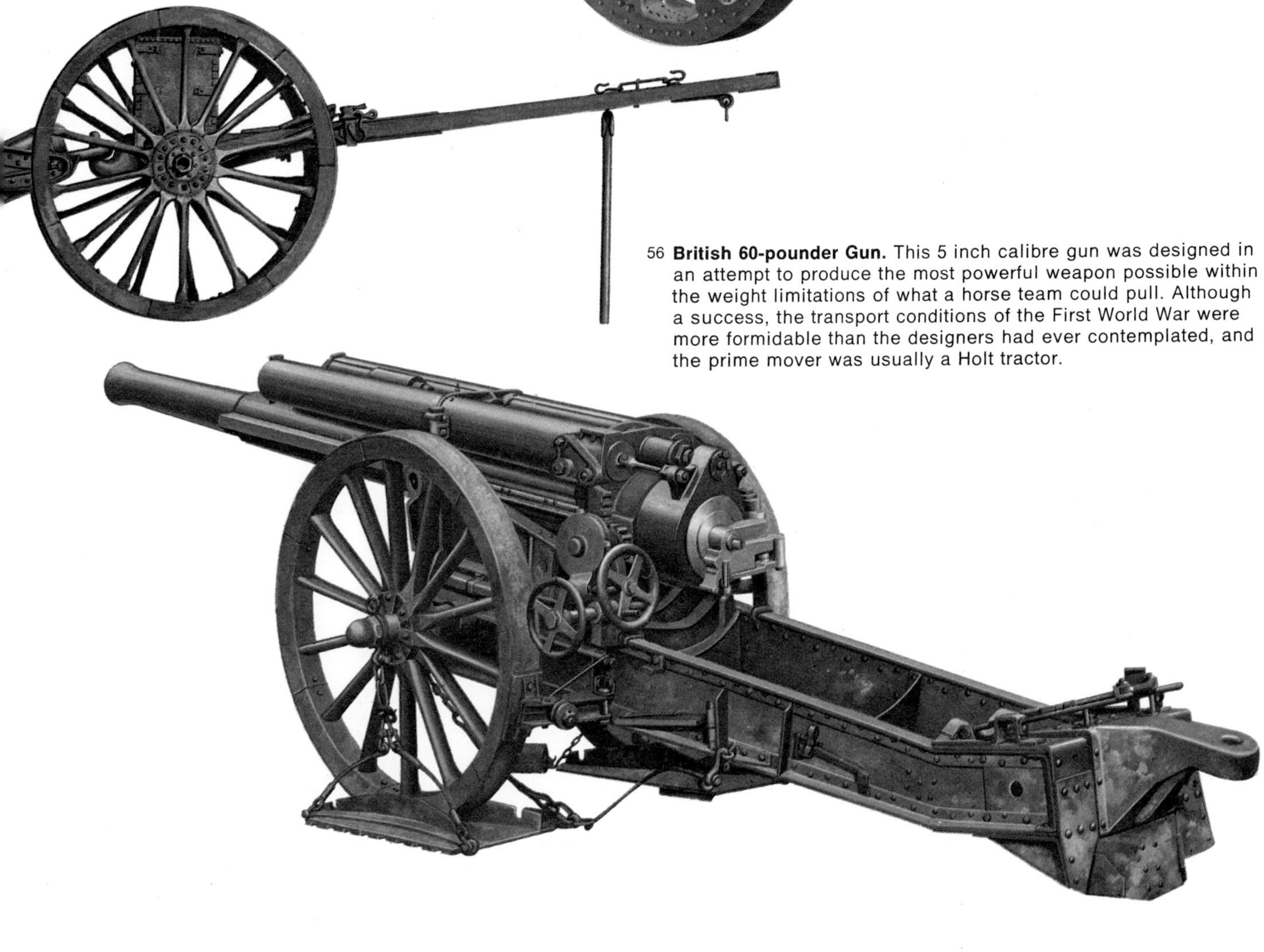

Lightness was a virtue where airborne operations were involved, and several elderly designs of pack and mountain guns were hastily revised, a notable one being the US 75mm howitzer. Although of the same calibre and firing the same shell as the M1897 gun, this was a vastly different design and one which could quickly be stripped down into seven parachute loads. As the war went on, techniques of parachute delivery improved and by the end of the war complete 25-pounder guns, 6-pounder anti-tank guns, and 75mm howitzers were being dropped. Probably the greatest contribution to airborne artillery firepower came from the German development of the recoilless gun, which we will explore elsewhere.

Another wartime demand was for flexibility of fire, and this reached its zenith with the German 105mm field howitzer designs of 1943. While the existing 105mm howitzer was satisfactory in most applications, experience in Russia led the German Army to demand a 105 which was less than the 4300 lbs weight of the existing gun, had a high enough velocity to give a good anti-tank performance, was capable of being concealed in forests and shooting from cover at high angles, and have a range of 13,000 metres without using rocket-boosted or other special ammunition, and finally, it had to be capable of 360 degrees of traverse because the Ivans had a habit of appearing from directions least expected.

This was a steep specification to meet, but Krupp and Skoda both produced weapons to fit. They looked similar, but the Skoda model was the better of the two. It used four trail legs to give support, allowing the gun to revolve completely around. With 80 degrees of elevation it could cope with any type of fire, and the long barrel ensured a worthwhile velocity and range. Only the end of the war prevented it from going into production, and it was closely studied by ordnance engineers of many countries afterwards.

When the war ended, it was 1919 all over again in one respect – the amount of surplus ordnance to be disposed of. The smaller countries of Europe were able to refit their armies with a choice selection of ex-German, British or American equipments, while those who fell into the hands of Russia were willy-nilly outfitted with Russian weapons.

57 **German 15cm M1916 Gun.** A wartime Krupp design, this could range to over thirteen miles and proved a thorn in the Allies' side throughout the rest of the war.

There was sufficient ammunition on hand to last a long time (Britain turned out 78,700,000 25-pounder high-explosive shells alone, during the war years, a figure vastly exceeding her expenditure) and when the Korean War erupted in 1950 it was fought entirely with Second World War equipment so far as artillery was concerned. It was not until the late 1950s that new designs began to be considered with any seriousness. Moreover, what money had been available since 1945 had been shovelled into the bottomless pit labelled "air defence", culminating in various involved missile programmes. As a result of the limited experience with missiles during and shortly after the war there was a considerable opinion that missiles were the artillery of the future and the gun was on the way out, and this idea not only took root in the air defence field – very properly as it turned out – but was inclined to spread into other artillery areas as well. But the necessity to equip for "brush-fire" wars took the edge off the argument, and field artillery has remained more or less faithful to the gun. How right this course was has been shown by naval experience – the US Navy equipped itself with guided missile vessels, and then found it had nothing suitable for carrying out an old-fashioned shore bombardment when the need arose in Vietnam.

Close support artillery today is largely controlled by the weight factor. Today, everything must be capable of being airlifted, either in a fixed-wing aircraft or slung beneath a helicopter, and this has had its effect on design. The new British 105mm light gun (81) is a good example. Large portions of it are made in high-strength light alloy; the trail sides are tubular in section to give maximum strength with minimum weight. The barrel can be slipped out and exchanged; two barrels are provided, one chambered for standard British 105mm ammunition as used in the Abbot SP gun and one for the NATO standardised US pattern ammunition: depending upon the theatre of operations and source of supply, the appropriate barrel is installed in minutes. There is no shield, nor any other waste weight, and the whole weapon can be helicopter lifted or towed behind the lightest truck. Withal, it ranges to 15,000 yards – an impressive performance.

What will come along tomorrow is anybody's guess. It will not contain any revolutionary ballistic principle – I doubt if there are any left to be discovered now. It will, I feel sure, be a logical progression, built on what has gone before, taking advantage of scientific advances in metallurgy and explosive chemistry to extract the utmost performance out of the least weight. But it will still have a breech and a muzzle and a trail and a top carriage – it will still look, sound, and behave like a gun – it will still have the same breed of slightly oily, slightly dirty, slightly deaf gunners clustered around it, who, at the close of the day's firing, will still slap the barrel affectionately as if it were a faithful horse. And when you talk of rockets or missiles, they will look at you pityingly and remind you of what Colonel Boxer, Superintendent of the Royal Laboratory at Woolwich, said in the 1860s: "Gentlemen, if rockets had been invented first, what a marvellous improvement we would have thought the gun to be."

Having considered what we might call the main stream, we can now take time to explore some of the branches and backwaters. Take, for example, one of the original purposes of artillery, i.e. the reduction of defences – siege artillery.

It was not until the middle seventeenth century, when fortresses began to be developed along scientific lines, that siege artillery began to take on a form of its own, and even then it was merely a heavier version of the standard smooth-bore gun of the day. Short-barrelled mortars firing heavy projectiles at high angle became the favoured siege weapon, largely directed towards searching behind cover to kill defenders, and smashing down the defences themselves, since while these were immensely strong against direct attack they were often very vulnerable to plunging fire. This rather haphazard technique was taken

58 **French 155mm Rimailho Howitzer.** Four of these were the equipment of each French Army corps in 1914, but they could hardly be considered a modern design, in spite of their very good hydro-pneumatic recoil system. They served through most of the war, however, until replaced by the 1917 Schneider design.

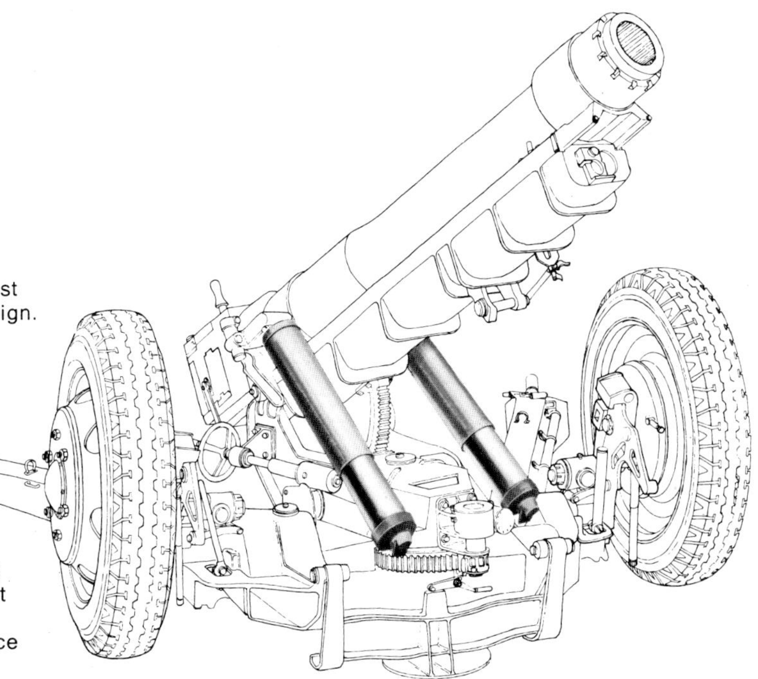

59 **Equilibrators.** Where the gun's weight is not evenly distributed about the trunnions, elevation and depression become difficult and an "equilibrator" is needed. This is the US 105mm M3 howitzer, and the presses push the cradle up to counterbalance the muzzle preponderance.

under examination in the 1840s by the French ballistician Piobert, who performed a series of experiments at Metz in 1844 and at Bapaume in 1847. Some old fortifications were to be demolished to make room for new, and Piobert mustered some artillery and did trial firings to find the most economical and efficient way of breaking down defences. He developed a scientific method of assault by direct fire, hammering the base of the wall until the upper portion collapsed of its own weight. In 1860 the British Army took up a similar experiment, using old Martello Towers as targets. The object here was to determine whether the new elongated shell fired from a rifled gun was any better at battering masonry than the smoothbore's spherical shell. They found that with a smoothbore it took 9684 lbs of shot and 3720 lbs of powder to reduce a tower to rubble, while the Armstrong gun did the same job with the expenditure of only 2953 lbs of shot and 511 lbs of powder.

From then on it was accepted that the smoothbore high-angle mortar had had its day, and that the direct-shooting gun was the siege weapon of the future. But this meant putting the gun out in the open, instead of concealing a mortar behind entrenchments, and many ingenious equipments were devised to allow the gunners some form of protection. Among the most ingenious was the Hydro-Pneumatic Carriage of Captain Moncrieff of the Edinburgh Militia. This officer had invented an ingenious carriage for coast artillery which disappeared, when fired, behind a parapet – as we shall later see. Now he applied his mind to the siege problem and supported a gun barrel on two arms hinged to the carriage. The barrel and arms were kept up in the air by a hydraulic piston and cylinder. To load, the gun was pulled down against the hydraulic pressure by a rope and tackle. After loading, it was released and the pressure forced the gun back up into the firing position (66). When fired, the recoil drove the barrel down again where it was caught and held in the recoiled position by a trip latch or a band brake. When loaded, the latch or brake was released and up it went again, poking its muzzle over the parapet. Thus the gunners were protected behind the parapet while loading, but the muzzle was clear of cover for firing. This is an interesting weapon if only for the fact that it was introduced into British service in 1870 and never appears to have been declared obsolete. It would be interesting to place a demand for spare parts, just to see what happens.

A similar device was adopted by the French for use in the other aspect of siege warfare; defence from inside the fortress. A 120mm gun on a similar mounting was placed on a rail truck so that it could be rapidly moved from place to place inside the defences, thus rendering it difficult to locate and hit and also enabling extra gun power to be brought to bear at threatened points (65). This was a timely reminder that the defenders might well have something to say about people who sat out in the open (even behind parapets) and attempted to knock the front door down, and it is noteworthy that the first major siege of the modern era virtually went back to the mortar. The Japanese at Port Arthur mustered eighteen 28cm howitzers (67) to do the battering, the first time such large weapons had been deployed in the field; a move which made Western armies think. Though their aim was poor – they once fired 260 rounds into the Russian fleet at anchor in the port for only six hits – they were invaluable as concrete-smashers.

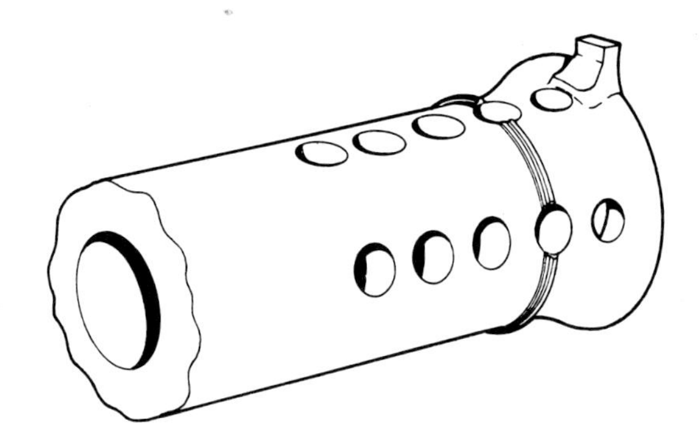

60 **An Early Attempt** to provide a muzzle brake, this was no more than some holes drilled in the end of the gun. At the time it did little good, but now that more is known about recoil behaviour, the idea is coming back into use.

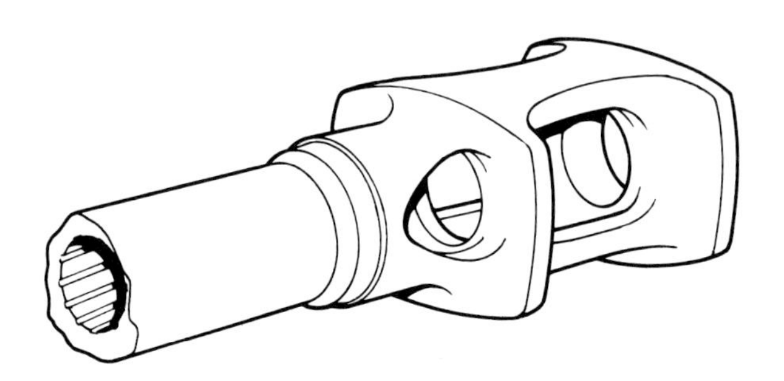

61 **A Wartime Design** from a German anti-tank gun, this four-port brake is of moderate efficiency without producing too much back-blast.

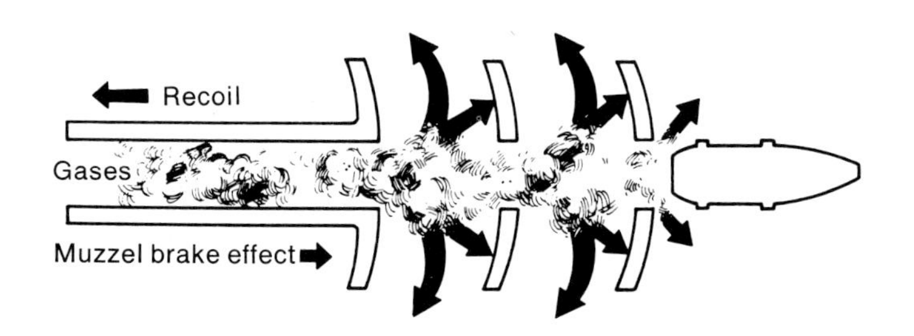

62 **How it Works** is shown in this diagram of a muzzle brake. The gas behind the exiting projectile is deflected so as to pull on the muzzle and thus reduce the recoil force.

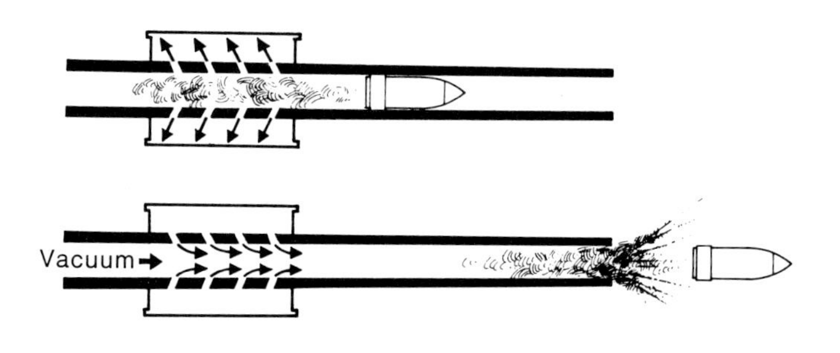

63 **Fume Extractor.** Used in tanks and self-propelled guns to prevent excessive fumes in the fighting compartment, the fume extractor is charged with gas at high pressure when the gun fires. As the pressure drops after shot ejection the trapped gas escapes, vented forwards due to the angle of the ports, and induces a flow of air as the breech is opened, thus exhausting any lingering fumes through the muzzle.

As we have already seen, the Germans and Austrians had already begun the development of the superheavies which were to flatten the Belgian and French forts in 1914, and the subsequent course of World War One showed that these heavy weapons could be used with normal HE shells to give support to troops in attacks by pulverising the rear areas to prevent reinforcements being brought up. From then on the superheavy gun (or howitzer) had an assured place on the battlefield, though there was a time during the 1930s when it was felt that they were no longer needed, that airplanes could do the job as effectively. This of course, depended on what you saw your Air Force as. Either, as Britain did, you saw it as an independent force, but then you still needed guns because the money was all going on fighters and long-range bombers, and close support of soldiers was something they had neither time, equipment nor inclination for. Or, as Germany did, you saw your Air Force as an extension of the Army, with ground strafing and dive bombing forming an integral part of the battle – and this was the better thinking. But in the 1930s the whole thing ultimately came up against reliability – would the airplane be there to drop the bomb when you shouted for it? And the answer was likely to be "it all depends on the weather". So the fact remained that in spite of all the assumptions and arguments, during the 1930s Britain and the USA did relatively little to develop bigger guns, while Germany spent a great deal of time and money in producing some exceptionally fine weapons.

Indeed, Britain entered the war in 1939 with precisely what she had left from the last one in 1919; a motley collection of 9.2 inch howitzers and 12 inch howitzers (75), enormous pieces which moved piecemeal and at a snail's pace, and took hours to assemble for action. They also had some 6 inch and 8 inch howitzers of similar vintage but slightly more mobile. The 6 inch was being replaced by a 5.5 inch of more modern design, but it took the 1940 Blitzkrieg to get rid of the rest. Most of the 8 inch and 9.2s were lost in France, which proved to be a blessing in disguise, since now something had to be done about replacing them. The 9.2 howitzer was too cumbersome for modern war, and a more mobile weapon was demanded and design began.

The 8 inch had too little range to be worthwhile in a fast-moving modern war, and a simple solution was to liner the barrel down to 7.2 inches, producing a range of 16,000 yards with a new 200 lb shell. The trouble was that this type of redesign did nothing to the awkward two-wheel carriage design. In spite of its hydro-pneumatic recoil system, the whole gun rolled back on firing, and to try and limit the recoil huge chocks were placed behind the gun wheels, so that when fired it could trundle up the chock and roll down again. Unfortunately, positioning the chock was a work of art; if you misjudged it, the gun either didn't roll back to the proper place or sailed over the top of the chock and crashed down behind.

The US Army, on the other hand, had seen the light a little earlier. They had purchased a 240mm howitzer from France in 1918, but it proved to be a bit of a lemon, and it was 1924 before the bugs were out of the design and it could be safely fired without blowing itself to pieces. Even so, it was not very good, and in 1939 work began on a scheme to develop a matched pair of weapons to fit a common basic mounting. One was a 240mm howitzer, to replace the earlier model, and the other an 8 inch gun to act as a long-range weapon. The barrels were developed

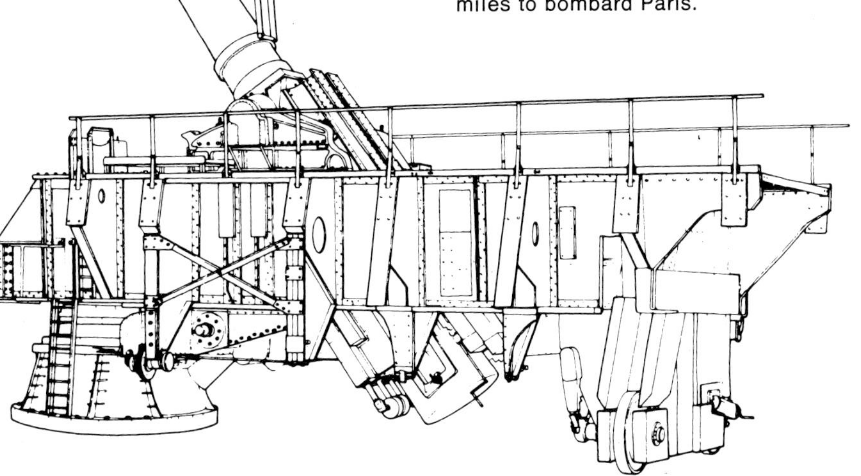

64 **The Paris Gun.** A 38cm gun lined down to 21cm, this weapon surprised the world in 1918 by throwing a 264lb shell over sixty miles to bombard Paris.

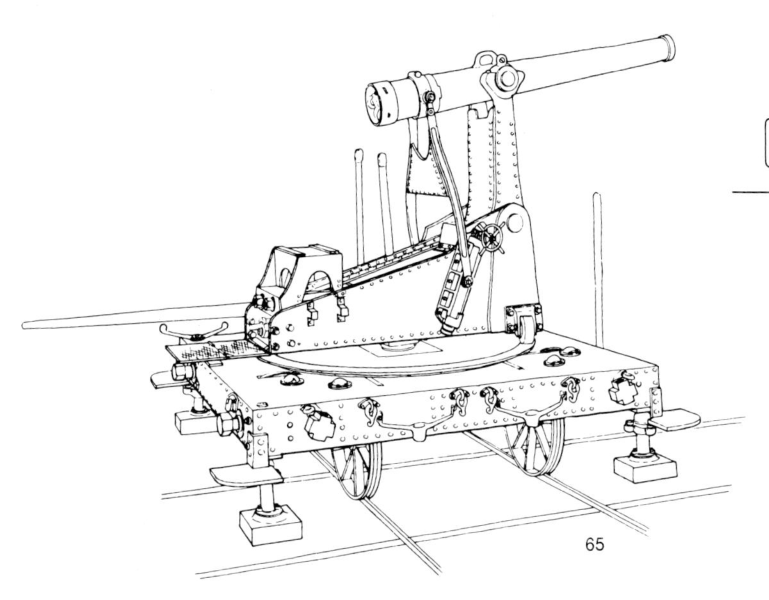

65

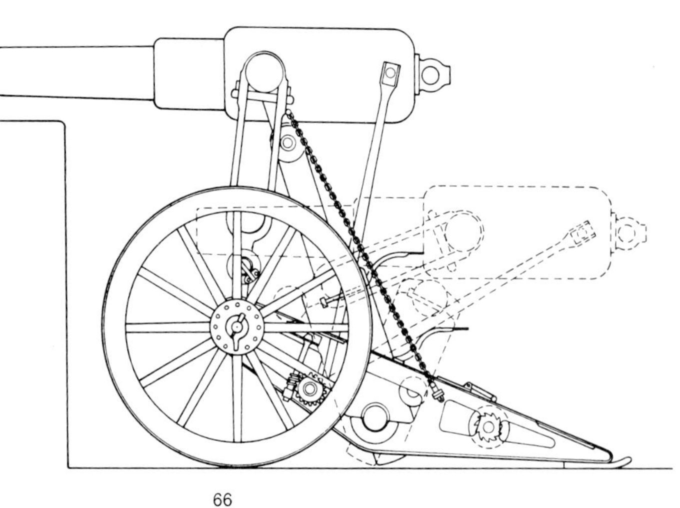

66

within a year, and by 1942 the equipments were being issued. As originally designed the mounting was built so as to form a semi-trailer unit behind a tractor, while the barrel (either one) was to be carried on a separate transporter, but after pilot models had been built to this design, they were abandoned in favour of a simpler system of carrying the carriage on a second transporter (76). Each unit was towed by a 38 ton Tractor M6, and a 20 ton crane formed part of the gun battery; this towed a trailer carrying a clam-shell bucket which was used to dig the gunpit. The crane then hoisted the carriage from its transporter and lowered it into position, and then lifted the barrel into its cradle on the carriage where it was secured by four large studs and nuts. The whole assembly could be done in less than an hour if the crane was available; without a crane the tractor winches could be used to bring the weapon into action, but this took a good deal more time. The two weapons were enormously successful, and, some having been supplied to Britain, the development of their improved 9.2 inch howitzer was dropped.

The US Army later began the development of another matched pair, this time a 240mm gun and a 280mm howitzer, but this was on a low priority, and as the 240mm/8 inch combination seemed to be doing all that was needed, the 240mm/280mm project languished. The carriage design was frankly copied from a German medium/heavy gun, the 17cm, which used an ingenious double-recoil system with the gun recoiling in the top carriage and the top carriage recoiling across the trail. This equipment was also furnished with a centre pivot platform onto which the centre of the carriage was lowered, while the rear end rested on a ground float. Small amounts of traverse were applied by operating a rack and pinion gear which moved the rear end across the float, pivoting around the central platform, but for large switches in direction the whole equipment was so well balanced that with the rack and pinion disengaged it could be freely swung around the central platform by one man – all twenty-seven tons of it.

This design so impressed the US technicians that they took it as their starting point for the 240/280 design. Soon the idea of a matched pair was dropped, and the project became one for a 280mm gun. Instead of splitting the

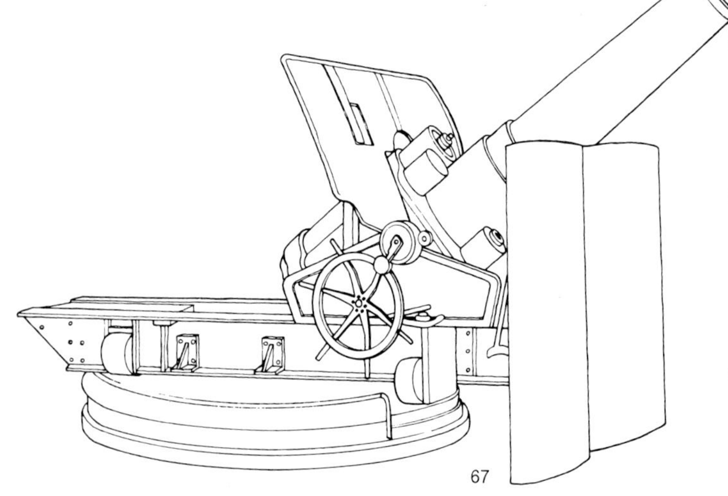

67

65 **French Fortress Gun.** A French adaptation of Moncrieff's idea, mounted on a rail trolley for rapid movement within a fortress system. Developed by M. Mougin, a director of the St. Chamond company, small numbers were made for the forts on the Franco-German frontier.

66 **Hydro-Pneumatic Siege Carriage.** One of Moncrieff's designs, the 6.6 inch Howitzer was intended for siege train duties. When fired the gun pivots the elevator arms back and forces the piston into its cylinder, compressing air. After loading in the 'down' position, a valve was released and the air pressure forced the gun back up into the firing position.

67 **Japanese 28cm Howitzer.** When the Japanese brought these weapons across country and emplaced them to bombard Port Arthur during the Russo-Japanese War, it was an eye-opener and a portent of things to come.

equipment and hauling the gun barrel separately, the American designers made use of their unrivalled expertise in heavy automative design and produced two special truck-tractors which simply hooked under each end of the carriage, hoisted it into the air, and drove off, with the gun still in place on the carriage.

This weapon entered the Hall of Fame as the first artillery piece to fire a nuclear projectile. But for all its ingenious design, it was still a cumbersome monster, and difficult to conceal. Once nuclear-armed surface-to-surface missiles and rockets became viable propositions, the superheavy gun was as good as dead, and once the US Army perfected Honest John, the writing was on the wall for the 280mm Atomic Annie (79).

The biggest calibre gun ever made was also an American wartime achievement. This was "Little David" (80), the 36 inch howitzer. An amazing weapon, it was a reversion to the Rifled Muzzle Loader concept. It was originally developed as a proving ground mortar for testing aerial bombs by lobbing them at targets, a simpler and more accurate method than by dropping them from aircraft. Some ingenious soldier saw possibilities, and the Army decided that a portable version might be of use in demolishing the strong fortifications expected on the Japanese mainland when Operation Olympus took effect. So Little David was modified to become a field weapon. The barrel was semi-trailed on a four-wheeled truck behind a tank transporter tractor, while another tractor dragged the mounting, a box 18 feet long, 9 feet wide and 10 feet high. A bulldozer dug a suitable pit, with ramps leading in and out, and the box was drawn into it, removed from its wheels and then packed around with earth to give a secure foundation. Then the barrel was pulled across the top and, with the aid of hydraulic jacks, lowered into its trunnion mountings.

The 3650 lb shell had a filling of 1550 lbs of HE and a pre-rifled driving band. The 216 lb charge was inserted

68 **Granny.** The Coventry Ordnance Works had already supplied the British Army with a 9.2 inch howitzer. After the appearance of the German monsters at the gates of Belgium they re-scaled the drawings and produced this 15 inch Model, which went into action manned by Royal Marines.

69 **Schlanke Emma.** The Austrian companion of Big Bertha, this 30.5cm howitzer was also unveiled in the Autumn of 1914 and showed the world that super-heavy mobile guns were not only possible but vitally necessary.

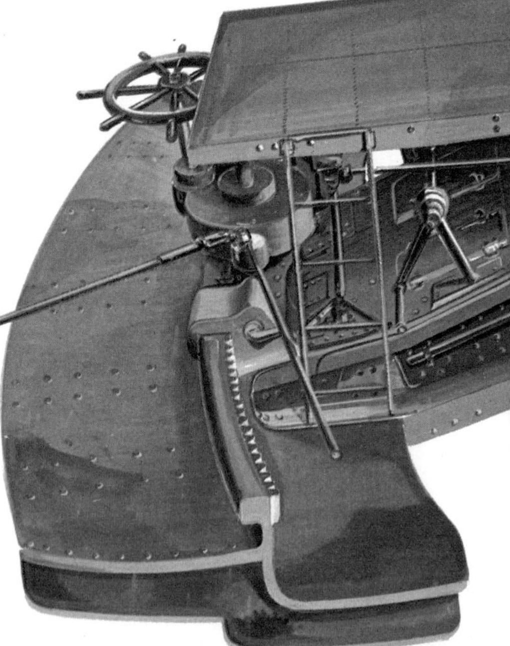

70 **Big Bertha.** Following the Japanese example, Krupp and the German Army developed this 42cm howitzer for the express purpose of demolishing concrete fortifications, a task at which they were soon seen to excel.

71 **US 8 inch Howitzer.** The M1 Howitzer was developed as a 'Partner Piece' to the 155mm gun, to fit on the same carriage. Firing a 200lb shell to 18,500 yards it served throughout the Second World War and remains in service today, its life extended by its nuclear capability.

into the muzzle and rammed, the shell was engaged with the rifling, and the gun elevated so that gravity slid the shell down into the chamber. The range was about six miles and the effect of the shell was devastating. But before the weapon could be shipped to the Pacific the war ended and Little David was relegated to the ranks of the obsolete.

But outranking all other siege artillery was the gun intended to demolish London in 1944. This was the fruition of an idea which has tempted many a designer and inventor; the multiple-chambered gun. One of the earliest attempts to make this idea work was the Lyman and Haskell gun tried at Frankford Arsenal in 1870 and again at Reading, Pa, in 1882. This was a six-inch gun, basically an Armstrong but with only three rifling grooves, and three splines on the shell. In addition to the usual chamber and cartridge, four more chambers were connected to the barrel. The basic charge was two pounds of black powder, and a seven-pound charge was placed in each of the auxiliary chambers. A wad behind the projectile formed the gas seal. The idea was that the basic charge would start the shell moving up the barrel; as it passed up the bore and uncovered the entrance to the first chamber, so

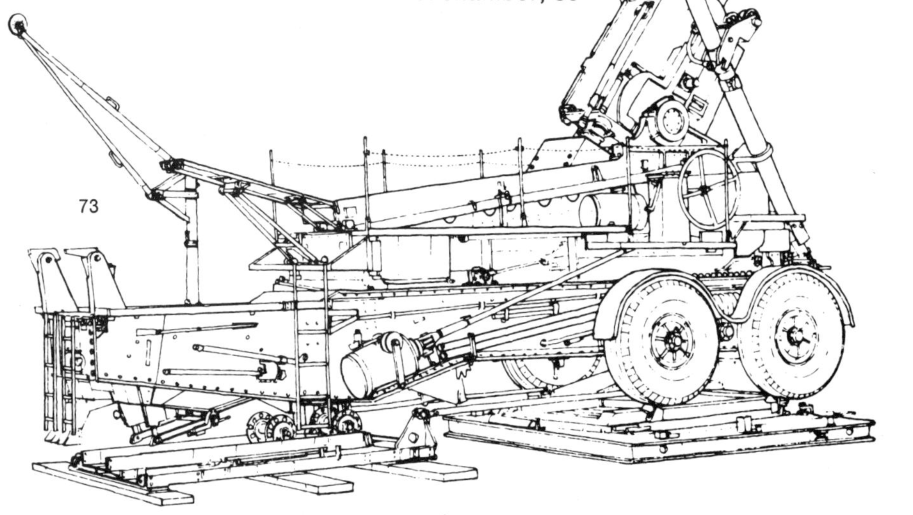

72 **Busy Lizzie.** Also known as the Millipede or V-3, this is the Conders multiple-chambered gun designed to shell London from Calais. As the shell is moved up the bore by the first cartridge, so the other charges are ignited in turn, giving the moving shell an additional boost.

73 **German 24cm Kanone Model 3.** A Rheinmetall design, though mostly built by Krupp, this was a modern weapon of outstanding design. With a 37.5 Kilometre range, nonetheless it was easy and simple to get into action.

74 An enormous 16 inch gun, is here being tested at Fort Tilden, military camp on Long Island, N.Y.

the flame would ignite the seven-pound charge which would explode and accelerate the projectile. In turn it would uncover the remaining chambers, each of which would fire its charge in turn, so eventually producing an exceptionally high velocity.

On trial it produced a velocity of only 1100 feet per second, which was much less than a standard Armstrong gun could manage, but with pressures in the barrel as high as 36 tons to the square inch, a figure which would make even a present-day ballistician think twice. The conclusion was reached that the flame from the base charge was getting past the wad and igniting the auxiliary charges before the shell got past them. The US Government was reluctant to back the design and the Lyman and Haskell gun died for want of funds.

Seventy years later, it appeared again, though it is open to doubt whether the new promoter had ever heard of Lyman and Haskell. The new man was Herr Conders, an engineer who was not an ordnance specialist but who had an idea. The idea was to build a very long gun with 28 side chambers arranged herring-bone fashion along the bore, working just the same way as the 1870 design. With a 15cm shell it was hoped to achieve well over 5000 feet per second muzzle velocity and a range of about 180 miles. By dint of political string-pulling he got Hitler's approval and work on the "High Pressure Pump" (or "Buzy Lizzie" or "The Millipede") began. A 2cm scale model seemed to work satisfactorily, and then a full-

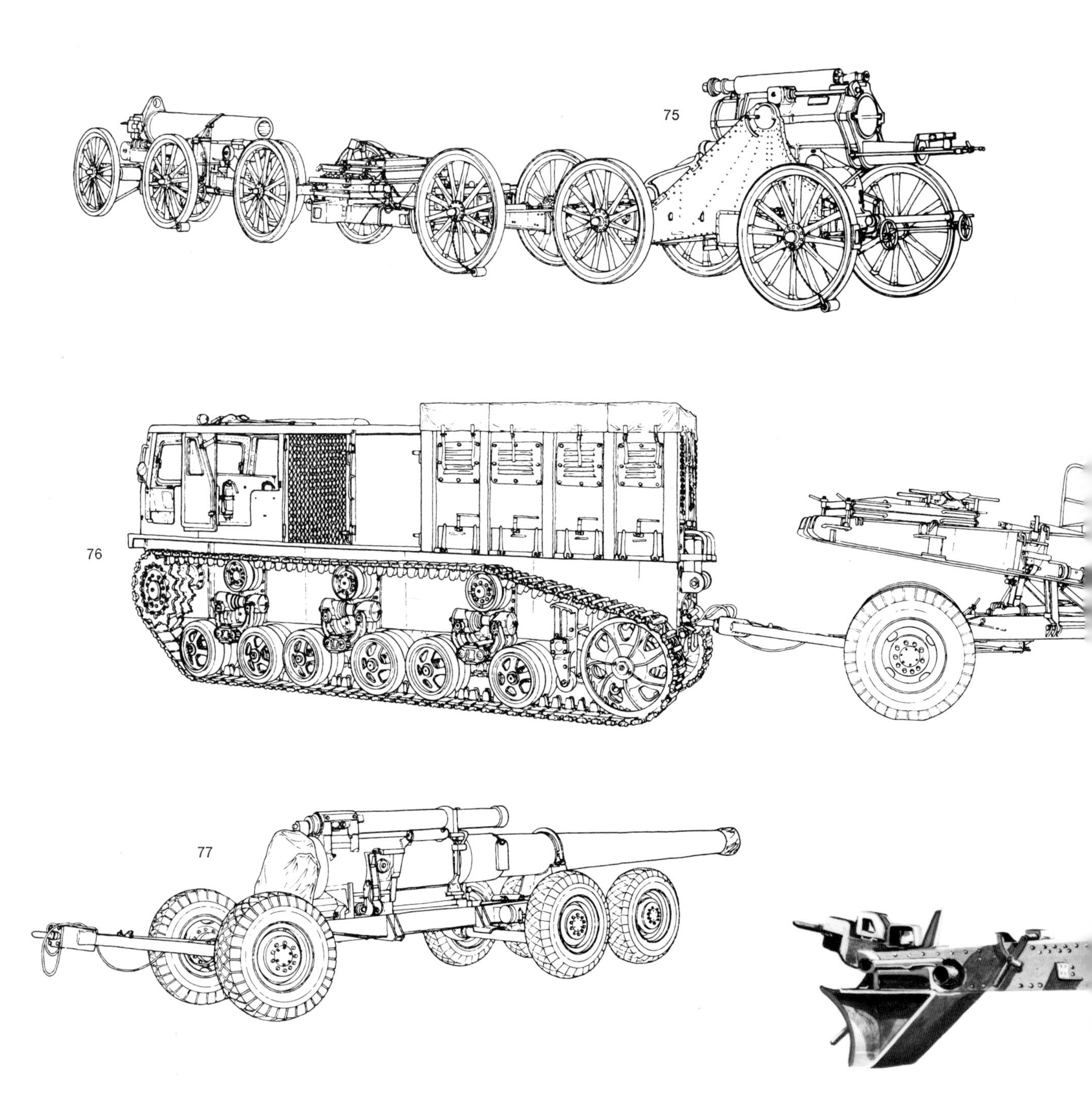

calibre short-barrel model was built at the Hillersleben Proving Ground. This was less successful, the shells proving unstable and erratic in flight. By now it was 1943, and Hitler, with various ideas about laying siege to London, christened the gun "Vengeance Weapon No 3" or "V-3", and ordered a fifty-barrel version, each barrel 150 meters long, to be built into a hillside near Calais (72).

Conders now built a full-size gun on an island in the Baltic and began training soldiers. But every time it fired a section of the barrel would blow out or some other misfortune would befall him. Eventually the Army Ordnance Office, who until now had been kept out of the picture by Hitler's express order, were called in to try and make it work. They succeeded in designing a good stable shell, but the sands had run out on Conders. The RAF saw the Calais installation, assumed it to be some sort of rocket-launching site, and bombed it to ruins. Two short-barrelled versions of the gun were built, one mounted on a railroad flatcar and one laid out on a hillside. Both were briefly used in the Ardennes offensive in late 1944, but they were blown up on the advance of the Allies, and Conder's dream finally lay in ruins.

But he had proved his point. If the Ordnance Office had been in on it from the start, there is every reason to believe that the Calais installation could have been in action by early 1944, and the last and greatest siege gun might well have laid London in ruins.

75 **British 9.2in Howitzer.** Moving the heavy howitzers was a problem in the early days of mechanical traction. They had to be divided into towable sections and reassembled on site. The 9.2 divided into barrel, top carriage and platform to travel.

76 **US 240mm Howitzer.** This shows the carriage on its trailer and the High-Speed Tractor M8 which was the standard towing vehicle.

77 **US 240mm Howitzer.** This weapon travelled in two pieces, barrel and carriage and this drawing shows the barrel, complete with ring cradle and recoil system, on its travelling wagon.

78 **British 5.5 inch Gun.** Standard medium gun of the British Army throughout the 1940s and 1950s; the upright 'horns' alongside the barrel are spring equilibrators.

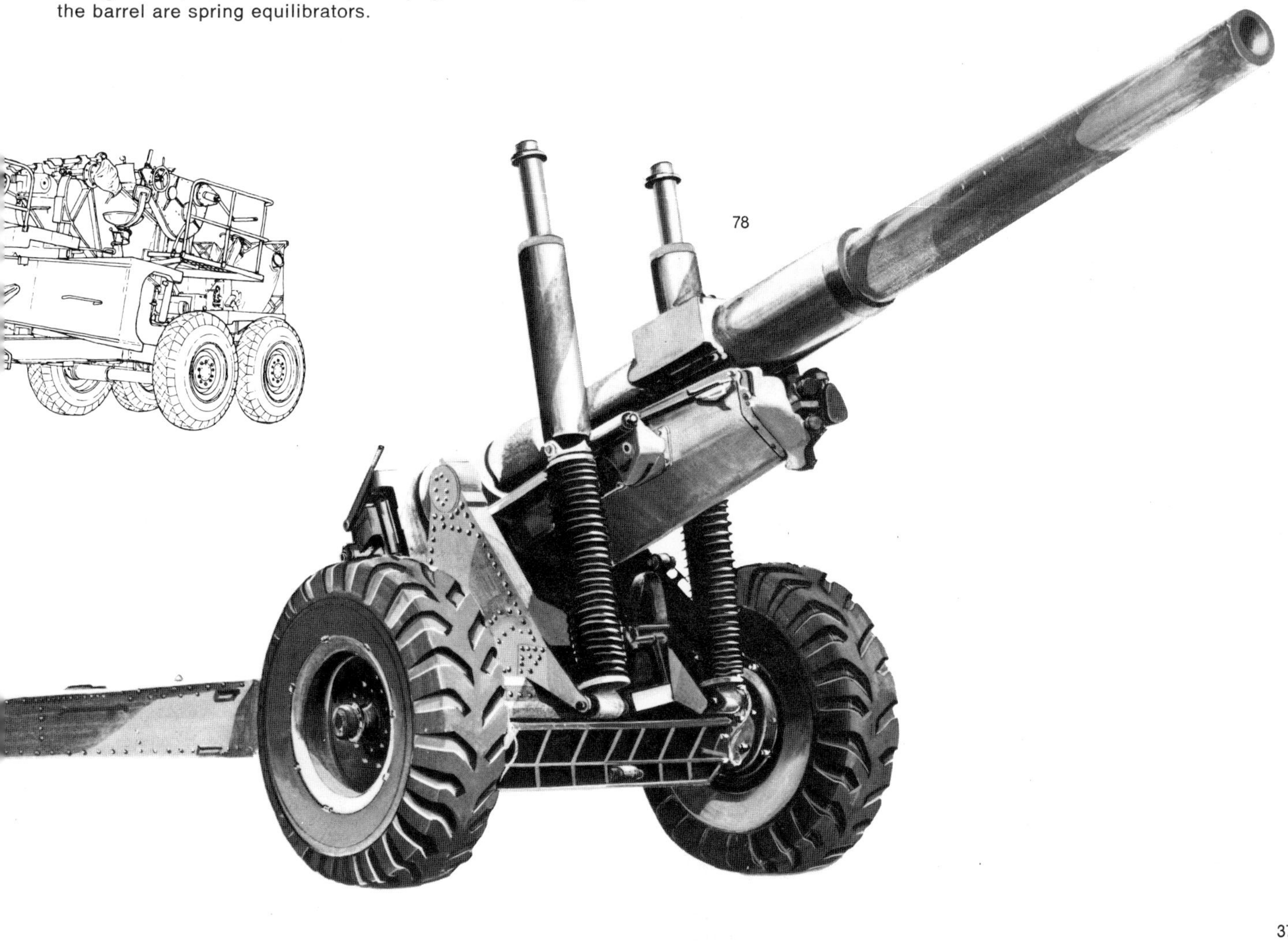

78

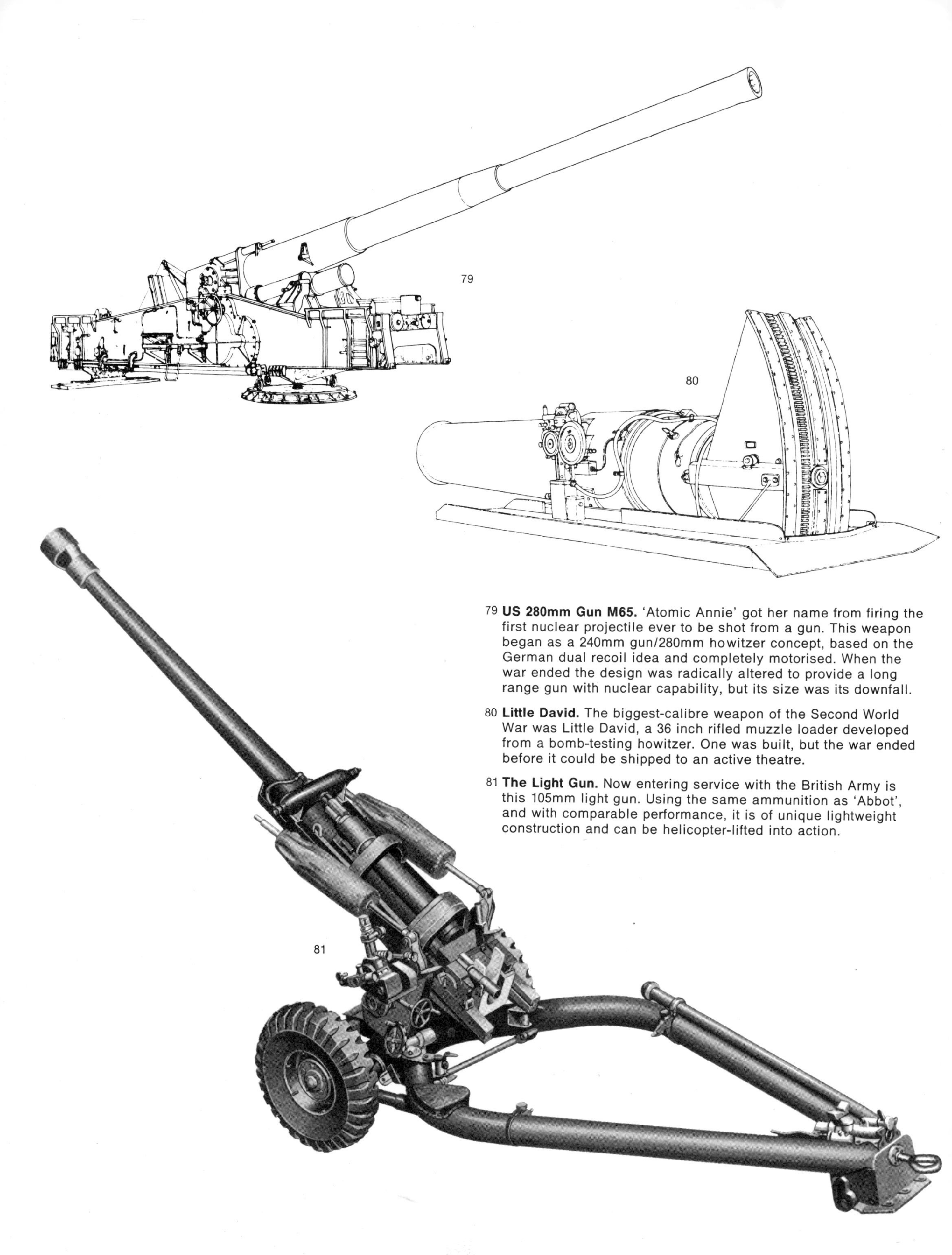

79 **US 280mm Gun M65.** 'Atomic Annie' got her name from firing the first nuclear projectile ever to be shot from a gun. This weapon began as a 240mm gun/280mm howitzer concept, based on the German dual recoil idea and completely motorised. When the war ended the design was radically altered to provide a long range gun with nuclear capability, but its size was its downfall.

80 **Little David.** The biggest-calibre weapon of the Second World War was Little David, a 36 inch rifled muzzle loader developed from a bomb-testing howitzer. One was built, but the war ended before it could be shipped to an active theatre.

81 **The Light Gun.** Now entering service with the British Army is this 105mm light gun. Using the same ammunition as 'Abbot', and with comparable performance, it is of unique lightweight construction and can be helicopter-lifted into action.

Rail guns

The first recorded amalgamation of the gun and the railroad appears to have been during the American Civil War (76) which is not altogether surprising. This was the first war in which rail communications played a vital part, and it was fought at a time when the rifled breech-loader was striving to gain a foothold and there was an interest in artillery which might otherwise have been absent. The first instance seems to have been the use of a 13 inch mortar on a flatcar in the campaign against Richmond, and one or two enterprising officers copied this idea in other actions.

But it appears that there had been thoughts about this very subject even earlier. In 1881-2, when the British Army landed in Egypt after bombarding Alexandria, Captain Fisher RN, a well-known gunnery man and future First Lord of the Admiralty, mounted some naval guns on rail cars and roughly armoured them. When reports of this were published in England, the Director of Artillery was surprised to receive a letter from a Mrs Anderson, who "submits a copy of a pamphlet on a system of National Defence by her late husband Mr J. Anderson, wherein it appears that so long ago as 1853 her husband proposed a complete plan of wrought-iron protected railway carriage. Mrs Anderson hopes that her husband's name will be associated with this subject if again brought forward, as appears to have been the case in Egypt." The Director of Artillery, for his part, "informs Mrs Anderson that no steps have been taken consequent upon her husband's proposals, and that the Secretary of State for War cannot admit that anything done in Egypt was due to them." Unfortunately I cannot trace a copy of the late Mr Anderson's pamphlet, but there is still a chance that he may have a legitimate claim to have originated the railway gun.

But simply flinging a gun on to the nearest flatcar doesn't make it a railway gun, though this is a trick which is often performed. Thus the Germans in 1916 mounted their normal 17cm field gun on a gondola (84). No, the railway gun is a specialist piece of equipment which has to be designed for the job in view. It has to fit the standard loading gauge so as to be able to be moved over the national rail network without complex re-routing. It must have carefully designed suspension so as to accommodate itself to hastily laid track in combat areas, track which would derail normal commercial rolling stock. And it has to be capable of firing from as simple a set-up as possible if it is to have any real tactical value.

This latter point may sound obscure but it can best be explained by looking at some representative equipments.

The general idea was toyed with mainly by Schneider of France during the opening years of the 20th century, but it was the First World War which brought the railroad gun into prominence. It is primarily a weapon of the continental power, who, with numerous frontiers to defend – or attack – requires its heavy reserve artillery to be highly mobile, and, given the static conditions of 1915-18 and the proliferation of rail feeder lines which sprang into being behind the two sides of the Western Front, the railroad gun was in its element.

It is difficult to pin down the first employer of railroad guns in France but it seems to have been the French who, in the early days of the war, found themselves short of heavy artillery. Manufacturers, arsenals, redundant forts and stockpiles were raided to provide gun barrels originally destined as reserve weapons for ships and fortresses. Since the mountings for these guns formed a permanent part of the ship or fort, the barrels so unearthed had to be given something to hold them off the ground, and Schneider, drawing on their pre-war experiments, rapidly designed some simple mountings. These were no more than massive boxes carrying the guns in trunnion mounts, without recoil systems and travelling on rail trucks. Once transported to their firing position they were jacked off the trucks and allowed to rest on the rails or on some form of foundation built alongside and between the track. When firing began the whole mounting and gun would then slide back a few inches at every shot. Direction was given by laying the track in a gentle curve so that positioning the gun on the curve gave coarse alignment, and fine pointing could be done by jacking the mounting to and fro. After a few shots the gun would have recoiled sufficiently to have moved out of alignment, and it was then hauled back into place by a hand winch anchored on the track ahead of the platform, or ,if the equipment was too heavy to winch, by jacking it back onto its trucks and hauling with a locomotive. Scores of these extempore weapons were built and fired; most of them eventually succumbed to cracking of the side-plates near the trunnions under the recoil stress, but in most cases this coincided with the barrel's being worn beyond repair, and the entire weapon would then be scrapped.

They served their purpose, though, while better weapons were being developed and produced, and by the time the war ended the French had produced a formidable collection of railway guns (85). The British also found that they could afford to utilise numbers of naval gun barrels produced for reserve stocks and proceeded to turn out rail mountings to suit. Among the first were the 9.2 inch guns which utilised the Vavasseur ex-broadside ship mounting. This was an inclined plane up which the gun recoiled, damped by hydraulic buffers and down which it ran back into battery by gravity (83). The geometry of this type of mount meant that the trunnions could be kept low so as to pass the loading gauge, but the original shipboard pattern was not suited to high angles of elevation. A modification was soon produced, being no more than a matter of wedging up the front end of the mount to give another 15 degrees, but this upset the delicate balance of the Vavasseur recoil arrangements, and the final solution was to build a special 9.2 inch gun with the breech area thickened and made heavier; this allowed the trunnions to be set farther back and still balance the gun, which in turn allowed more space behind the breech when the gun was elevated. Numbers of all models saw service in France and the special barrel version, the Mark 13, remained in service until the Second World War.

There are three basic designs of railway mount: the

82 **Civil War Innovation.** This cannon on a flatcar has a good claim to being father to all the railroad guns. Used during the US Civil War it was simply a standard siege piece tied down to the car bed and with a simple shield, but is sowed a productive seed in many minds.

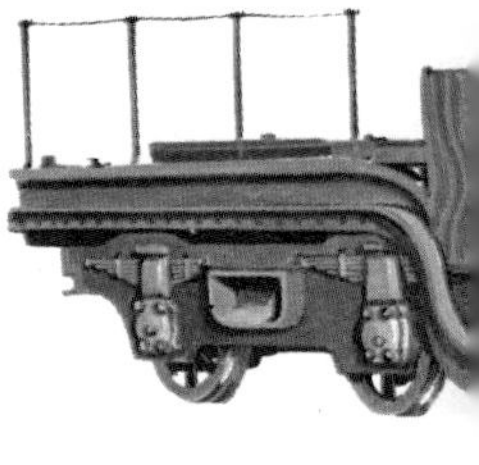

sliding mount, which the French pioneered; the rolling mount, in which the gun rests on its wheels and rolls back to absorb some of the recoil, the rest being taken up by a normal recoil system; and the outrigger mount, where the wheels are lifted from the track and the weight and firing shocks taken by outriggers and jacks. The latter is usual in calibres up to about 8 inches, since it is capable of absorbing shock from any direction and the gun can be fitted to a base ring on the mounting and traversed and fired in any direction. But railway guns of small calibre are barely worth the trouble and most of the service types have been of the rolling pattern, with more or less fanciful modifications.

The US Army were enthusiastic for railroad guns. Since they entered the war in 1917 with practically no heavy artillery they were driven, like the French, to stripping out spare coast guns and producing rail mounts to suit. After satisfying initial demands, they sat down to develop some better designs, since they had become attracted to the idea of using railroad guns for coast defence.

Under the impetus of war all sorts of inventors and theorists come bouncing to the front. One theory which was prominently touted in US military journals in 1917-18 was the Luellen-Dawson system in which reinforced bases were to be prepared and built on rail lines around the coast of the United States and on which special rail mounted guns in armoured cupolas could be positioned when the need arose (91). This sort of system sounds good until you work out the cost. It is comparable to a similar idea which was put forward in England at about the same time, a proposition to mount a 15 inch gun every 30 miles along

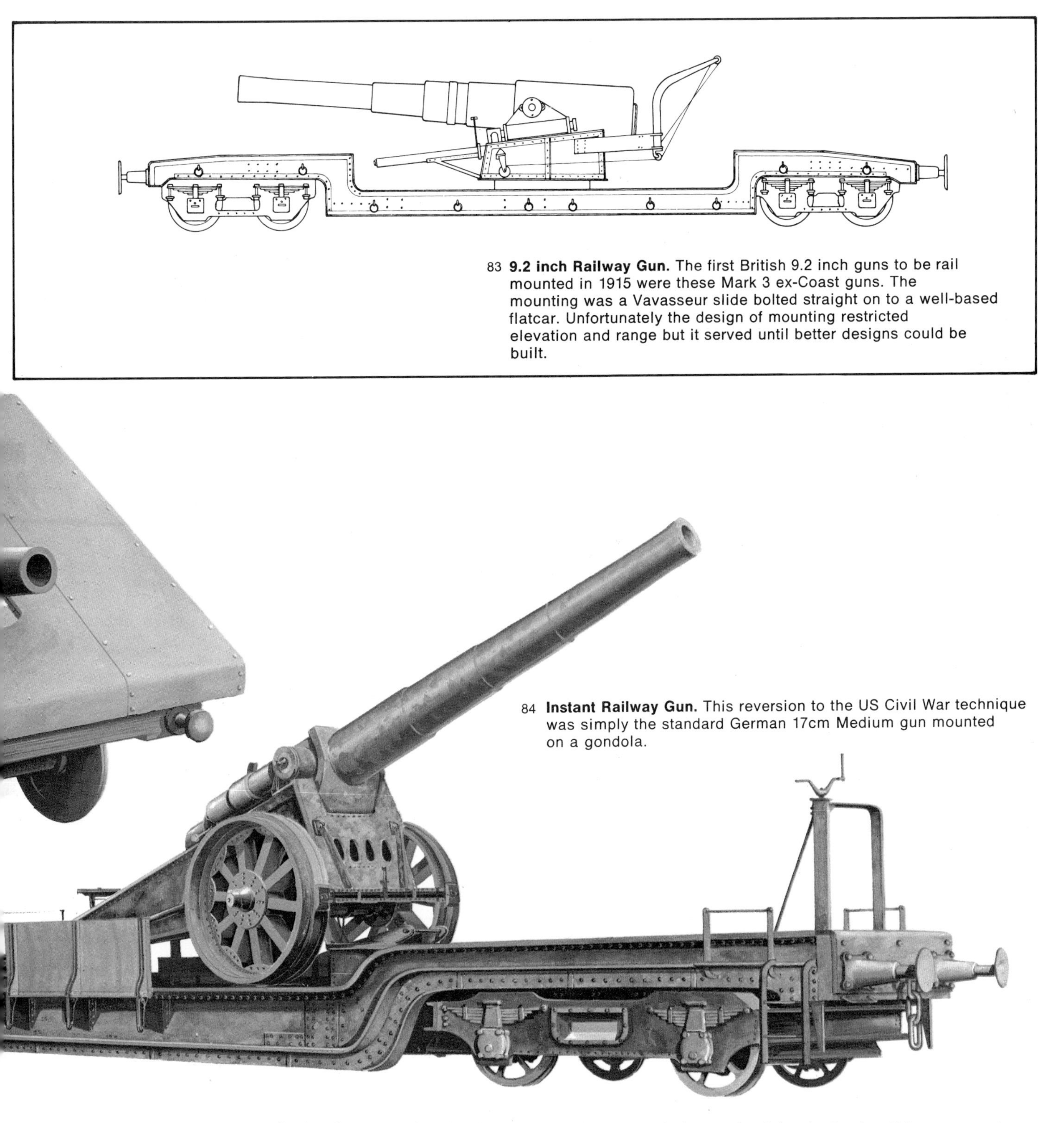

83 **9.2 inch Railway Gun.** The first British 9.2 inch guns to be rail mounted in 1915 were these Mark 3 ex-Coast guns. The mounting was a Vavasseur slide bolted straight on to a well-based flatcar. Unfortunately the design of mounting restricted elevation and range but it served until better designs could be built.

84 **Instant Railway Gun.** This reversion to the US Civil War technique was simply the standard German 17cm Medium gun mounted on a gondola.

the entire British coast. Outlandish as the Luellen-Dawson system was, it nevertheless contained a grain or two of wisdom, and the possibility of using railroad guns as a mobile coast artillery reserve was appealing. Some excellent designs resulted from this idea; firstly the US Navy produced a streamlined 14 inch gun mount as an experiment, one at least of which was sent to France in time for the final days of the war (90). A better design was the Army model 14 inch which used a mount copied and adapted from a French original (93). In this the gun was so mounted that it could be laid low in the chassis for travelling, and then raised by jacks for firing so as to allow space beneath the breech to permit recoil at high elevations. Additionally a turntable base ring could be placed on the ground, the gun mount lowered on to it and the trucks removed, the weapon thus becoming a semi-permanently emplaced gun with an all-round traverse.

Another design was a 16 inch howitzer (94) which was also intended as a coast defence weapon for delivering plunging fire on to ships' decks, but for a variety of reasons. finance among them, the design was not put into service.

The Germans, as befitted a country with a dense rail net-

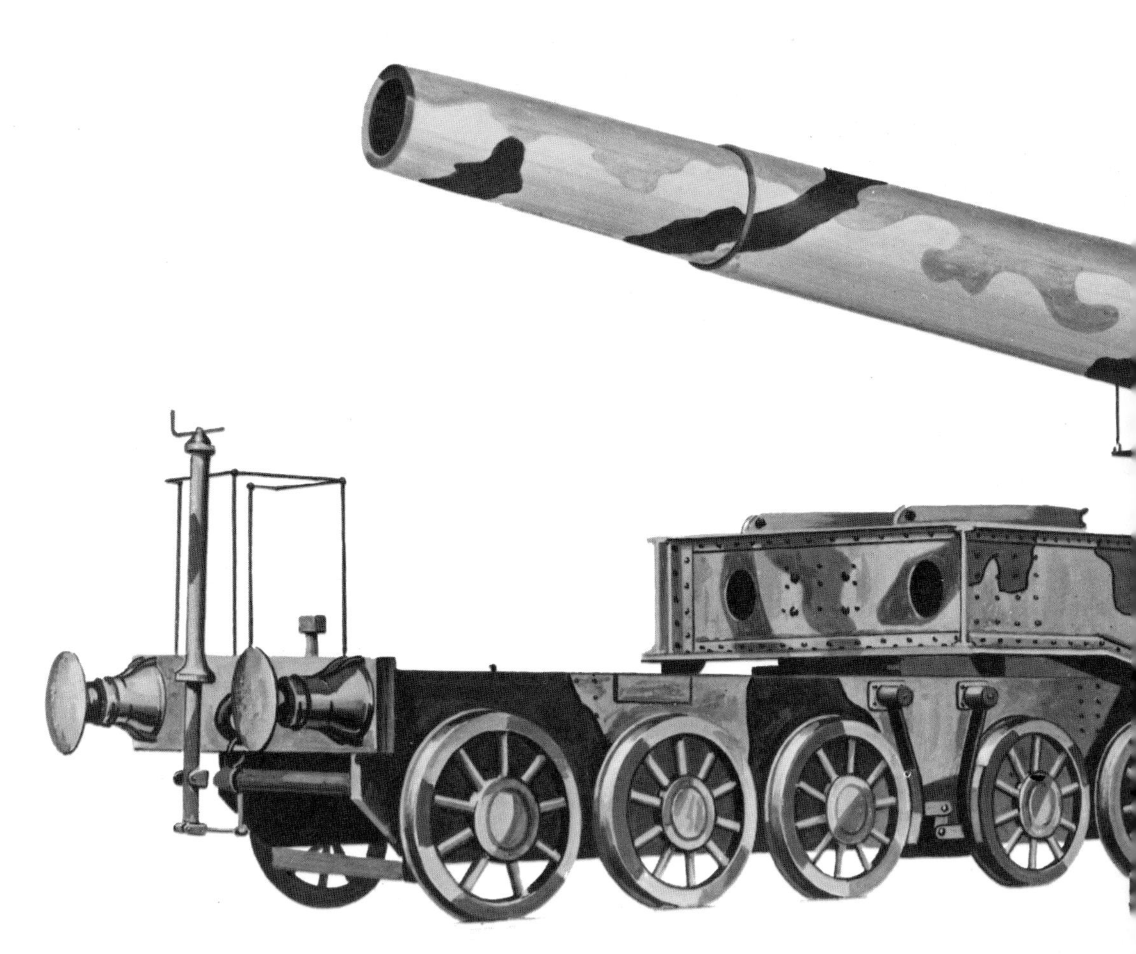

85 **French Railway Gun.** A 40cm gun on a railway mount, the weight is taken from the spring suspension by jacks prior to firing.

work and many frontiers, also took up railroad gun design and produced several models, of which the 38cm Max (87) was one of the best. This was often emplaced near the Belgian coast and used to bombard Allied positions at long range. Their new technique of pulling the gun back into a nearby railroad tunnel after every few shots had been fired handed the Allies a fine problem in discovering who was shooting at them, with what and from where.

The rail gun which caught the imagination of the world in the First World War was, of course, the Paris Gun (64). Developed by Rausenberger of Krupp's, this was a conventional gun taken to its logical conclusion. By taking a 38cm Max as a starting point, linering down the barrel to 24 centimetres and adding some twenty feet to the length, it became possible to propel the shell at a sufficient velocity to take it into the stratosphere. Here the air resistance was negligible, and in consequence the enormous range of 82 miles was possible. Mark you, this wasn't achieved without some drawbacks making themselves apparent. Every time the gun fired the erosion was so severe as to move the shot seating some distance up the bore and thus increase the chamber volume; to counter this the projectiles were serially numbered for each barrel, with driving bands of gradually increasing diameter. Moreover the chamber volume was measured after each shell had been rammed home so that a correct weight of charge could be calculated and weighed out and bagged for that shot, so as to maintain the correct muzzle velocity. Viewed as an effective weapon of war, in the balance of effort, time and money expended versus damage done, the Paris Gun was a gross waste, but as a propaganda weapon it had immense value and this was not forgotten. Another feature of the Paris Gun which the German Army didn't forget was that it had been inspired and manned by the Navy, and before the war was far behind them, they were preparing plans to produce a better weapon "next time". With Hitler's accession to power they got the go-ahead and began developing the 21cm K-12 Gun as the long-range gun of the future; with the assistance of Krupp's the ballisticians had decided on a rifling system very similar to that of the old RML guns. A barrel with eight deep grooves was made, and the shells were provided with eight curved ribs to travel in the grooves, with a copper sealing ring at the rear. The barrel was 105 feet long and weighed 98 tons and, like the Paris Gun, was braced to withstand the firing whip. Due to the muzzle-heaviness of this vast length, the gun mounting had to be designed so as to allow it to be jacked up one metre so that the barrel could be balanced and leave enough room for the breech to recoil without hitting the railroad track beneath. The second gun built was provided with powerful hydro-pneumatic balancing presses so that the trunnions could be farther back,

removing the need to jack up and down.

Lengths of track were carried on the gun train, and when going into action a T-shaped spur was laid, along the "upright" of which the gun was pushed until the front truck came to the junction with the cross-piece. Then the mount was jacked up and the truck turned through 90 degrees and jacked down again so that the truck wheels engaged with the trail of the cross-piece. Pointing of the gun was now done by motoring the front truck right or left, pulling the gun around the pivot of the rear truck. Once laid the trucks were clamped firmly to the track and a dual recoil system took the shock, the gun and top carriage recoiling along the mounting and the mounting recoiling across the truck bolsters.

The 21cm K-12 served its purpose insofar as with a range of 93 miles it out-ranged the Paris Gun and put the Navy firmly in its place, and as a technical exercise it taught a lot of valuable lessons. But as a practical weapon of war it was of less value. It appears to have been used once or twice in 1940 to shell England across the Channel and there are reports of shell fragments found in Kent at that time which correspond to the dimensions of the pre-rifled projectile. Beyond that employment – which did no damage – it seems never to have been used, and once the system had been proved to work no more were made.

An equally pointless weapon, from the practical point of view, was the enormous "Dora Gerat" 80cm railroad gun designed by Krupp in 1937. For many years it has been believed that two of these monsters were built, nicknamed "Dora" and "Gustav", only one of which was ever used. Recent researches however indicate that only one equipment existed. Krupp's named it "Gustav" in honour of Gustav von Bohlen und Krupp, but the German gunners were an irreligious lot and preferred to give it their own name – "Dora" – and so the misunderstanding arose. The whole equipment, in firing order, weighed 1350 tons and moved piecemeal to its firing point. There a dual track was laid and the gun mount built up with the aid of an overhead crane, a process which took the best part of six weeks. 1420 men under a Major-General served to operate, assemble, maintain and guard the weapon. Two projectiles were provided, a four-ton high-explosive shell and a seven-ton concrete-piercing shell, with maximum ranges of 29 miles and 13 miles respectively.

After proof firings in Hitler's presence at the Rugenwald range in 1942, the gun was sent to add its weight to the German attack on Sevastopol where it fired about 40 rounds. The only other record of its appearance is at Warsaw during the 1944 rising where it again fired some 30 rounds. After this, it disappears from view, and this disappearance is a minor mystery of the war. If, as has been suggested, it fell into Soviet hands, it is remarkable that it

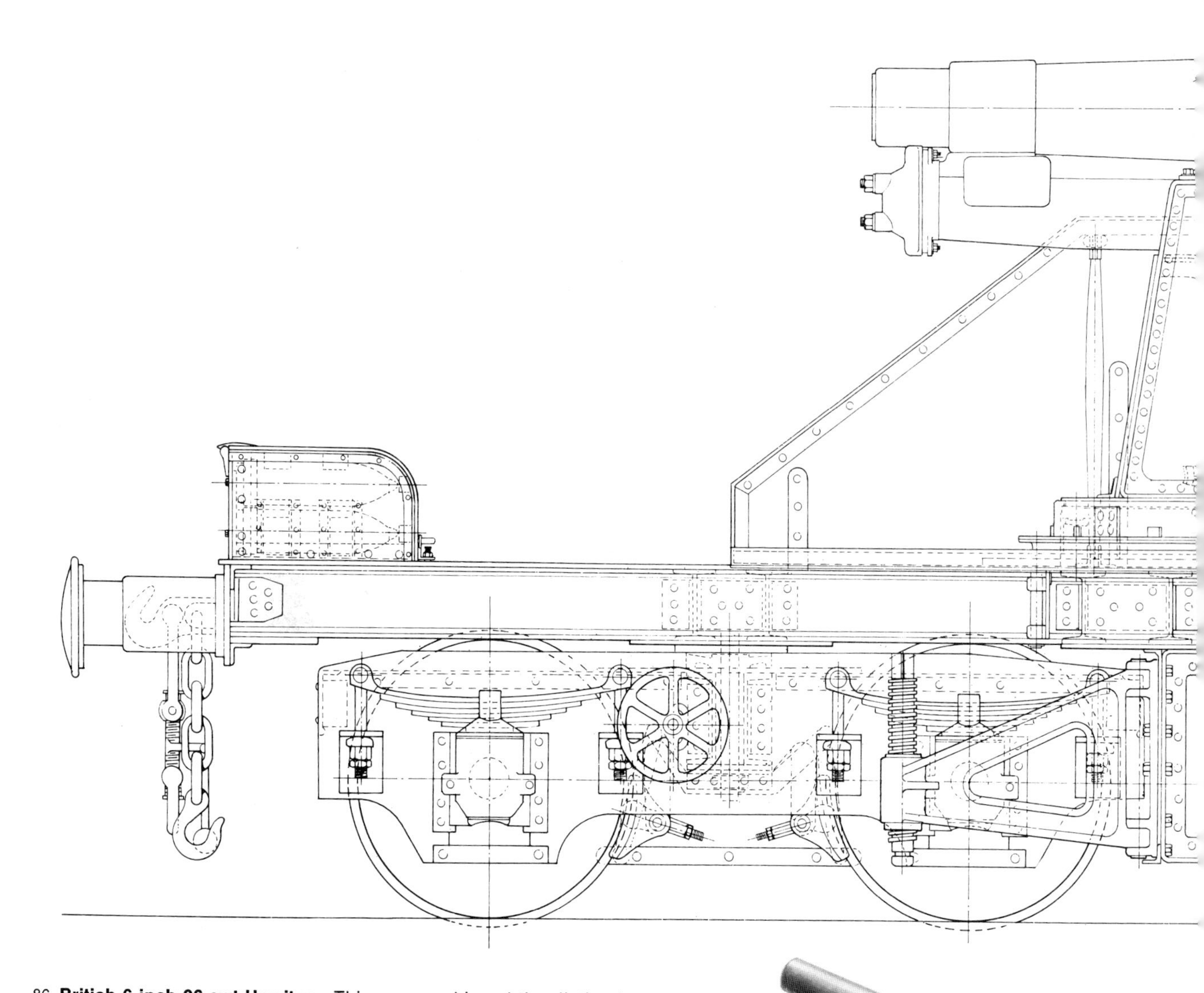

86 **British 6 inch 26 cwt Howitzer.** This never achieved the distinction of official adoption, but was a British design to give mobility to the standard medium howitzer of World War One. With a range of 11,400 yards the rail network of the Western Front could get it close enough, but the 100lb shell was hardly worth the trouble, and heavier weapons became the accepted rule for rail mounting.

87 **Max E.** The 38cm Max E Railway gun, used, among other tasks, for the German bombardment of Verdun in 1916. In this drawing it has been removed from its wheels and emplaced on a turntable.

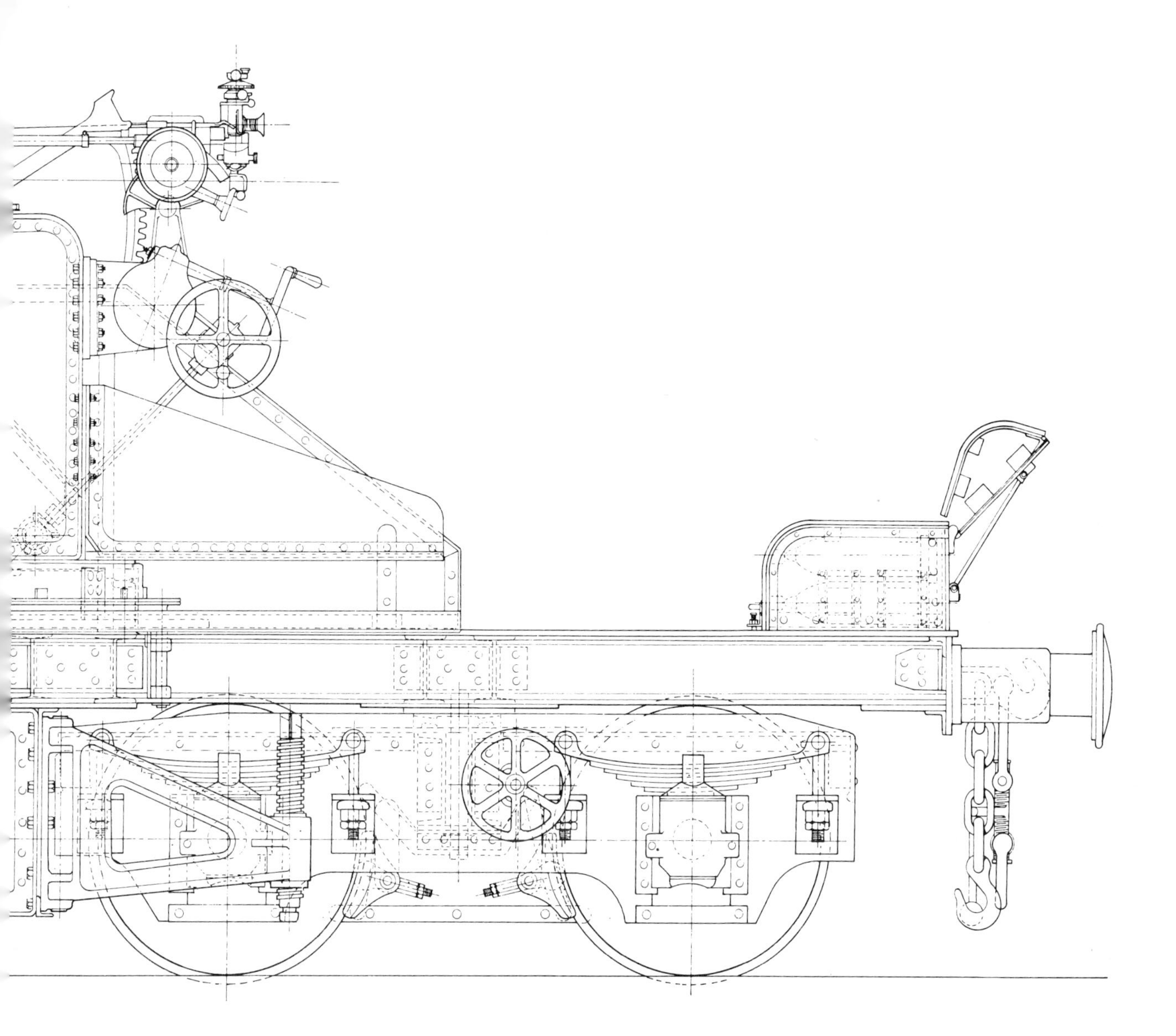

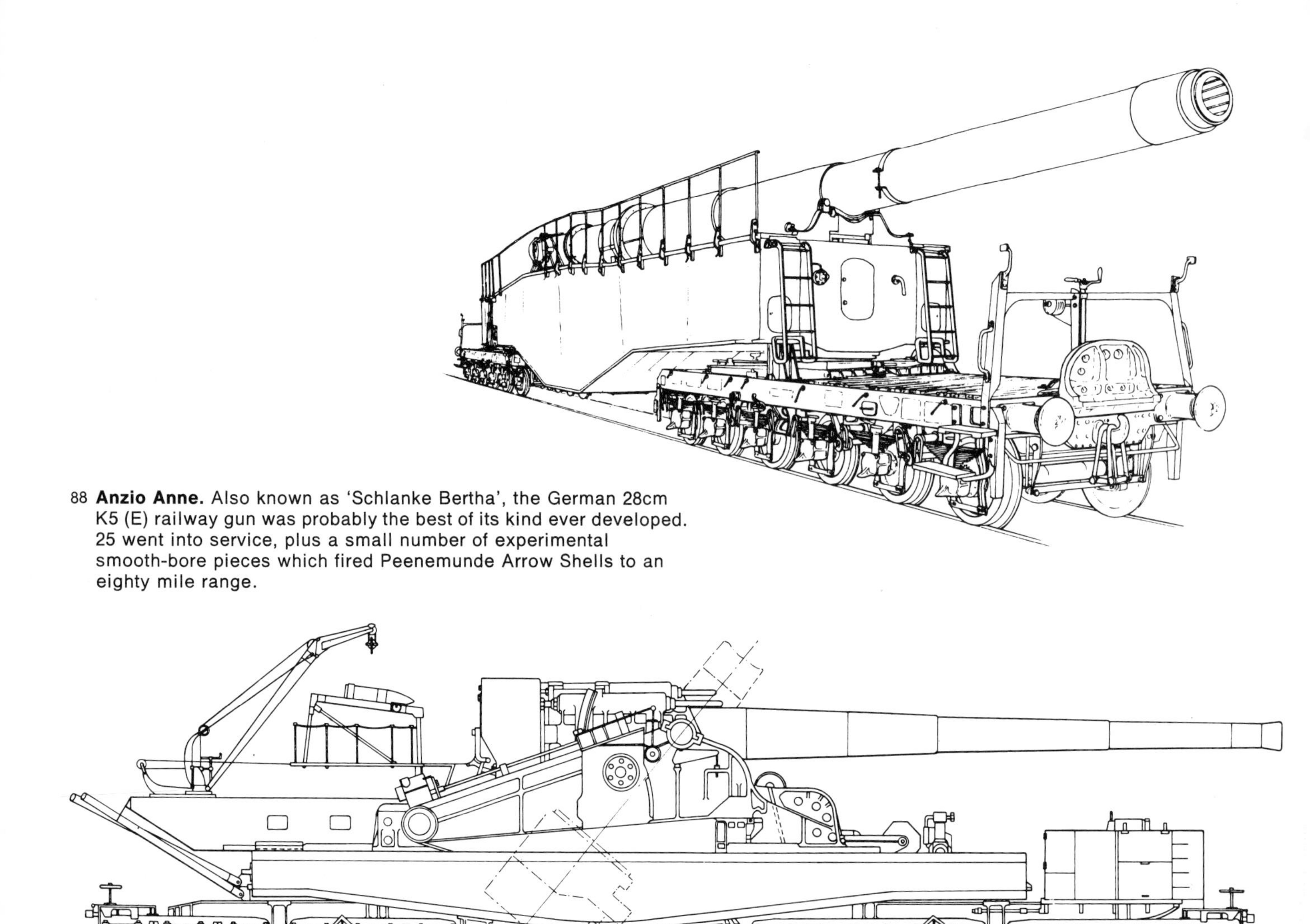

88 **Anzio Anne.** Also known as 'Schlanke Bertha', the German 28cm K5 (E) railway gun was probably the best of its kind ever developed. 25 went into service, plus a small number of experimental smooth-bore pieces which fired Peenemunde Arrow Shells to an eighty mile range.

89 **The Batignolles Mount.** This French design of mounting was adopted by the US for their 14 inch gun to allow them to keep the gun low for travelling and then to jack up the top carriage to give room for the breech to recoil at high elevations. Several were built in the 1920's and deployed through the USA as mobile coast defence guns.

is not on display somewhere; this was such a monster weapon that surely any Allied nation who captured it would have made a considerable propaganda virtue of it and installed it in some museum or display. Yet the only traces of it found when Germany was occupied after the war were a spare barrel and some rounds of ammunition at Krupp's proving ground.

By far the best railroad gun ever developed by the Germans – or anyone else for that matter – was the 28cm K5, known to the German Army as "Schlanke Bertha" and to the Allies in Italy as "Anzio Annie" (88). This, using the research done on the K12, was made with a deep-rifled 12-groove barrel to fire a 561 lb splined shell to 68,000 yards (almost 39 miles). In order to better this, a rocket-assisted shell was developed; this held a solid-propellant rocket motor in the head with a blast pipe running down through the shell body to the base, with the high-explosive payload packed around the pipe – a pretty problem in insulation. A time fuze in the nose ignited the motor when the shell was at the top of its trajectory and the additional rocket boost extended the range to 94,000 yards (53 miles). Unfortunately the payload was so reduced by the rocket motor and blast pipe that the effect at the target was hardly worth the effort, and, like all rocket-boosted shells, its accuracy was poor. This is due to the fact that a shell is rarely perfectly aligned on its theoretical trajectory and may be "yawing" some few degrees in any direction at any given instant. Thus, at the instant the rocket motor thrust takes effect, if the shell is yawing 'up'", the range will be increased, while if it is yawing "down" it will be decreased. Similarly, sideways yaw will throw it off course. Due to this the area within which the K5's rocket shell might fall was some 5 kilometers long and three wide, which is hardly pin-point shooting.

Not satisfied with this increase in range, the gun was now bored out to 31cm smoothbore and a heavier cartridge designed. The projectile for this version was designed at the Rocket Research Establishment at Peenemunde, using

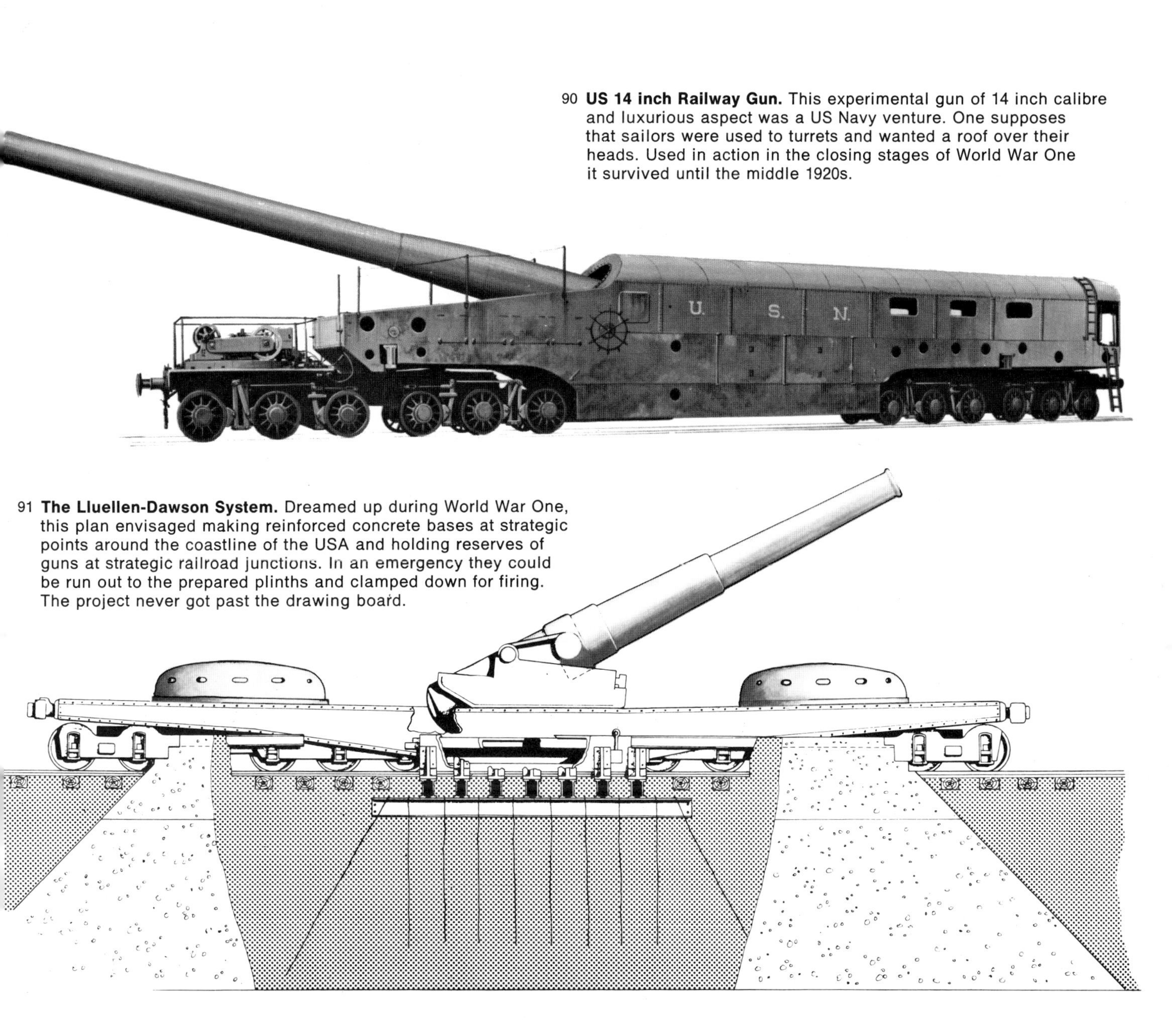

90 **US 14 inch Railway Gun.** This experimental gun of 14 inch calibre and luxurious aspect was a US Navy venture. One supposes that sailors were used to turrets and wanted a roof over their heads. Used in action in the closing stages of World War One it survived until the middle 1920s.

91 **The Lluellen-Dawson System.** Dreamed up during World War One, this plan envisaged making reinforced concrete bases at strategic points around the coastline of the USA and holding reserves of guns at strategic railroad junctions. In an emergency they could be run out to the prepared plinths and clamped down for firing. The project never got past the drawing board.

the wind tunnels which had seen the development of the V-1 and V-2 missiles. It was known as the "Peenemunde Arrow Shell" and was a 70 inch long slender dart of 12cm calibre, with four fins and a central belt of full 31cm calibre. When fired with the special charge it left the bore at well over 5000 feet per second, the belt was discarded, and the streamlined dart whistled off to a range of 160,000 yards (slightly more than 90 miles). Again the payload in the slender projectile was negligible, but it was a notable achievement, and two 31cm guns were issued with a supply of Arrow Shells, which saw action in the closing stages of the war. Some 28 of the standard version were built and were used in all theatres of war at various times.

So far as the rest of the world was concerned, the railroad gun had had its day. The veterans of the First World War which had been greased and tucked away in 1918 were dragged out and cleaned in 1939, but apart from a few which were spread around the south coast of England in 1940-41, very few could be said to have gone to war. The British 18 inch howitzer and 14 inch gun (93) were sited near Dover where they could join with the coast guns to duel with the German batteries on the far side of the Channel and thicken up the coast defences, and a battery of 9.2s went to France in 1939, but they got left behind in the rush in 1940. In 1944 eight 9.2 inch guns and the two 18 inch howitzers were made ready to go over the Channel to provide some weight to deal with anticipated fortifications on the German border, but they were not required. When the war in Europe ended in 1945 the air forces had shown that when it came to delivering large quantities of high explosive at long range they were a reasonable, if not quite so accurate, substitute, and the railway guns of the world were scrapped. A 28cm K5 from Italy was taken across to the USA for evaluation and now sits, the sole survivor of its species, in the Aberdeen Proving Ground. It is the last of the dinosaurs.

92 **British 12 inch Rail Gun.** Another Elswick design using ex-Naval guns from HMS 'Cornwallis'. Two others were built by Vickers, and are recognisable by having the wheels entirely exposed.

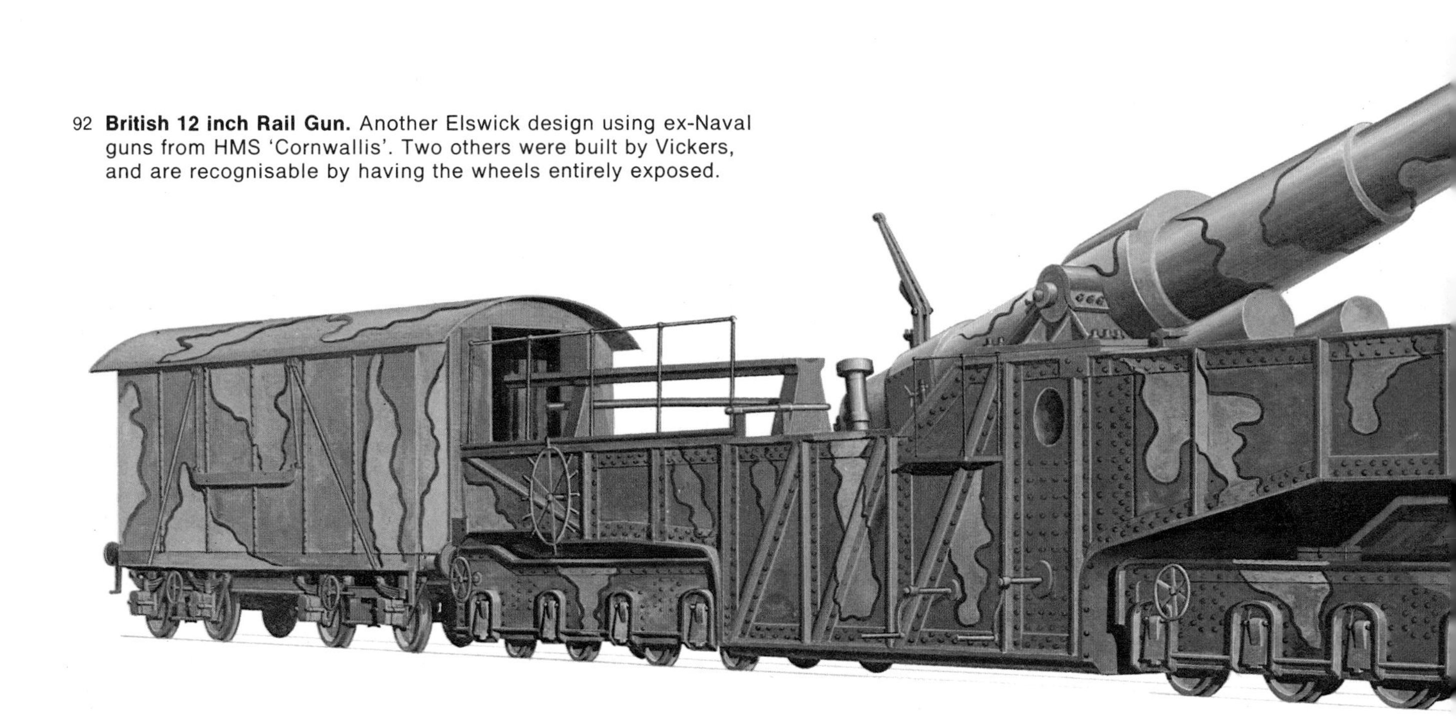

93 **British 14 inch Rail Gun.** In 1916 the Elswick Ordnance Company had two guns for Japan which could not be delivered. Offered to the Army, they were taken into service as railroad guns known as 'His Majesty's Guns "Boche-Buster" and "Scene-Shifter" '. A simple but effective design they fired a 1400lb shell to 38,000 yards.

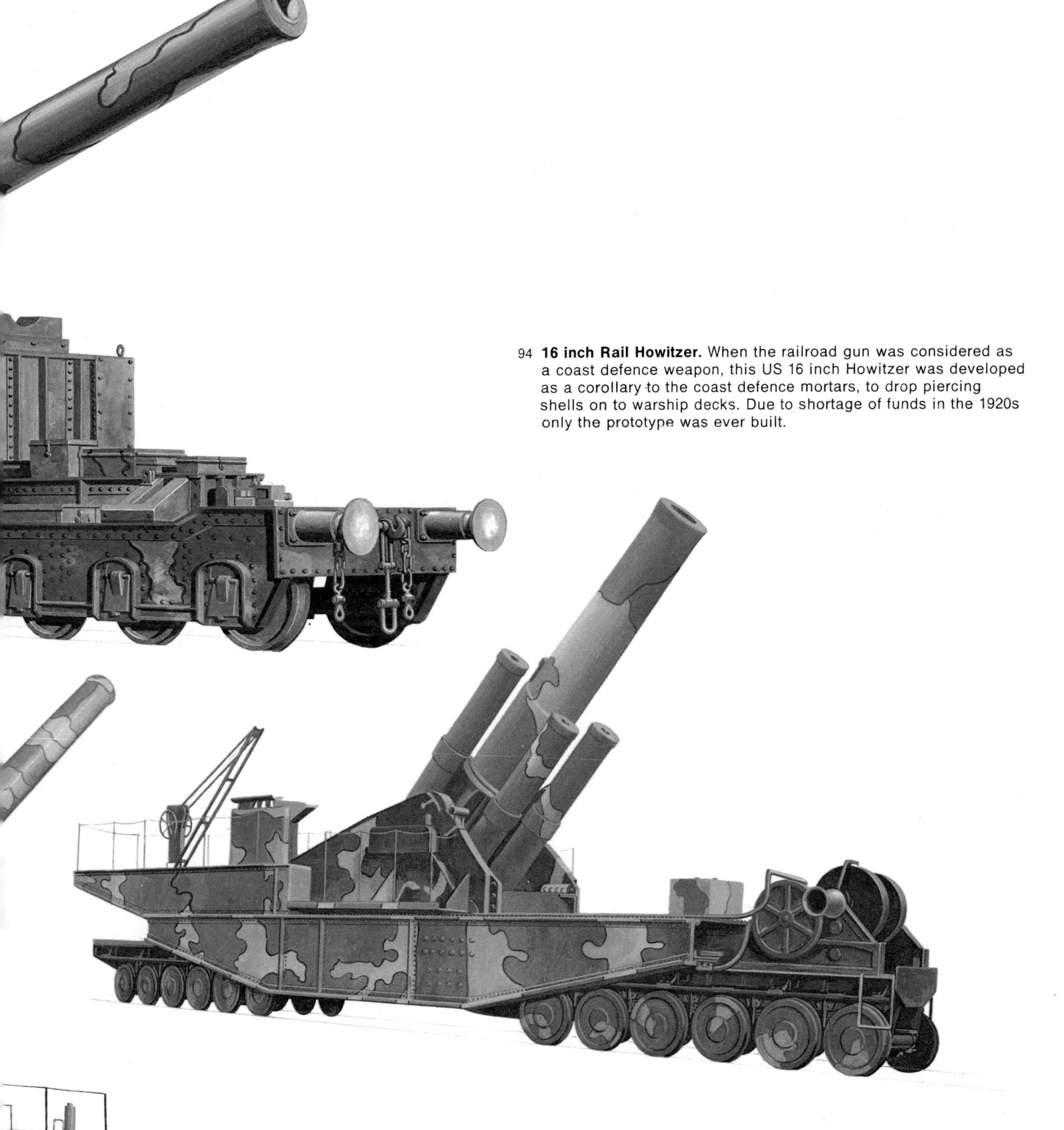

94 **16 inch Rail Howitzer.** When the railroad gun was considered as a coast defence weapon, this US 16 inch Howitzer was developed as a corollary to the coast defence mortars, to drop piercing shells on to warship decks. Due to shortage of funds in the 1920s only the prototype was ever built.

Coastal guns

People had been putting guns on seafronts with a vague intention of shooting at invaders almost ever since the gun was invented, but the rise of Coast Artillery proper did not begin until the middle of the nineteenth century. Two things gave impetus to this: the rise of the iron ship, and the Crimean War. In the latter campaign the combined British and French fleets made several attempts to attack forts at Sebastopol with indifferent success, and the efficiency of the Russian defence led many other nations to ponder. The iron ship, no longer very vulnerable to round shot fired from smoothbore guns, also led military staffs to consider the frightful possibility of raids on their naval dockyards by enemy fleets of such vessels. Almost every nation with a seaboard began to look seriously at the possibility of attack and defence against it. In Britain, Lord Palmerston ordered a Commission to investigate the state of defences in 1859 and to make recommendations to improve matters. They duly delivered their report in 1863 and a vast construction programme began in Britain covering every major dockyard and port. These works took many years to construct, being finished about 1880. Over the years the armament was frequently changed, as new weapons came along, and one can honestly say that they were never finished, revisions of armament and improvements in ships always left them a bit behind the times.

Other nations put their houses in order in similar fashion. Germany and France armed their coasts and in the USA in 1873 the "French" Board met to discuss the protection of the enormous coastline; among other items they proposed an expenditure of no less than 5½ million pounds sterling for the defence of San Francisco, which some critics felt to be a little lavish in view of the geographical layout of the area. In 1885 the "Endicott" Board looked into the matter again, and this time the money flowed like water. As Cuba, Panama and the Philippine Islands came into the American sphere, so fortifications had to be produced to protect them, as in similar fashion Britain was called upon to provide forts and guns for almost every corner of her Empire.

At the time of the great revival of interest in coast defence, the breech-loading gun was making its appearance. The Armstrong gun was touted as a useful weapon, and in Germany several of Krupp's larger breech-loaders were installed. But Britain decided that the Armstrong was unsuited to coast defence, simply because the breech closing was insufficient to withstand the pressures demanded. The object of the coast gun was quite firmly

95

A Railroad gun is an overpowering beast close up. Any gun is an awesome thing when fired at night. The combination of both—in this case a French railroad gun firing at night—eloquently displays the terrible majesty of the Great Guns in full voice.

Photo: ECP Armees.

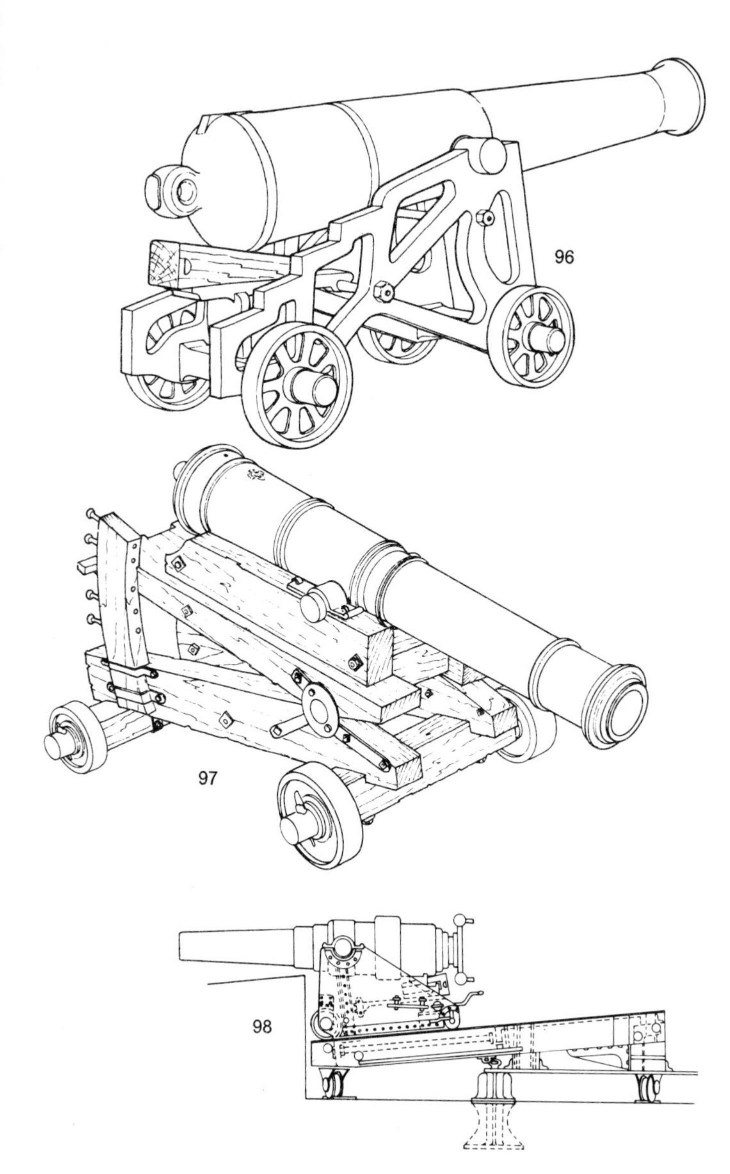

96 **The Garrison Standing Carriage.** Much the same as the Naval truck carriage of the same era, the Garrison Standing Carriage is simply a rectangular frame to support the gun with wheels to allow it to recoil.

97 **Kohler's Depression Carriage.** Now celebrated by the badge of 22nd (Gibraltar) Battery, Royal Artillery, this is the carriage invented to defend the Rock at close range during the siege of 1779-83. It was the first carriage specifically designed to fire downwards.

98 **Garrison Sliding Carriage.** In order to control recoil the standing carriage was placed on an inclined plane or 'slide'. This was mounted on trucks to allow it to be traversed from side to side so as to cover a greater field of fire.

99 **Moncrieff's First Model.** The first successful disappearing carriage this used a sub-carriage to keep the gun at a constant angle as it disappeared. The counterweight, at the foot of the rolling arms, was an iron box filled with several hundredweights of gravel.

100 **Moncrieff's Mark Two.** The improved Moncrieff carriage did away with the sub-carriage and had the counterweight made of massive cast iron blocks.

101 **The Buffington Carriage.** The original Buffington design of counterpoise carriage, about 1860. While theoretically sound it would have been difficult to make up into a operational weapon, and it had to wait for improvements before it proved its worth.

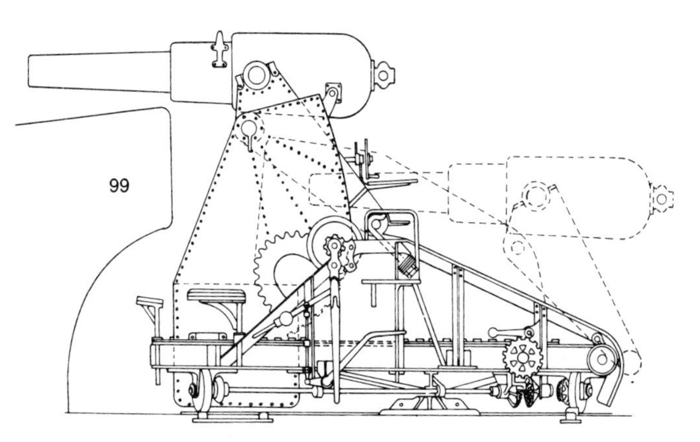

recognised – it had to smash through enemy armoured vessels. To do this the shell had to have as much velocity as possible and this demanded high pressures and powerful charges, both incompatible with the primitive breech closing systems of the day. This was the principal reason behind the reversion to muzzle-loading, though many Armstrongs and smoothbores were retained for flank defence and landward defence of the forts.

The standard carriage was the simple "Garrison Standing Carriage" (96), the soldier's name for what the Navy called a "ship carriage". This had been in use for innumerable years, being permitted to roll back on recoil and then be heaved back with handspikes. An improvement in the early years of the 19th century was the "Sliding Carriage" (98) in which the ship carriage, without wheels, was placed on an inclined plane, up which it could slide. But with the arrival of the shrapnel shell it was necessary to produce something which would allow the gunners rather more protection against enemy fire, and which would if possible conceal the gun except when it was actually firing, thus giving the seagoing gunners nothing to aim at until it was too late.

As far back as 1835 a Colonel DeRussy of the American Army had suggested mounting a gun on a form of standing carriage in which the wheels were mounted eccentrically, so that as it rolled back it would descend behind the fort's parapet for concealment. On the face of it a fairly sound idea, but one which was difficult to put into practice, largely due to the difficulty of running the gun back into position. Then in the 1860s Captain Moncrieff of the Edinburgh Militia proposed a form of counterbalanced gun carriage. In 1871 twenty of these were issued. The gun, a 7 inch RML, was mounted in a light carriage and this was connected to the top of a pair of curved arms which could roll back on a lower carriage. The rear of the gun carriage had two wheels which ran in an inclined plane. Thus when the gun was fired the recoil forced it back so that the curved arm rolled and allowed the wheels to run down the plane, bringing the gun below the level of the parapet. A large counterweight was fitted to the bottom of the curved arms, to resist the force of recoil, and this together with the curved face of the arm brought the gun to rest where it was retained by a pawl. Loading was now carried out under cover and then the pawl was released allowing the counterweight to swing the gun up over the parapet into the firing position (99).

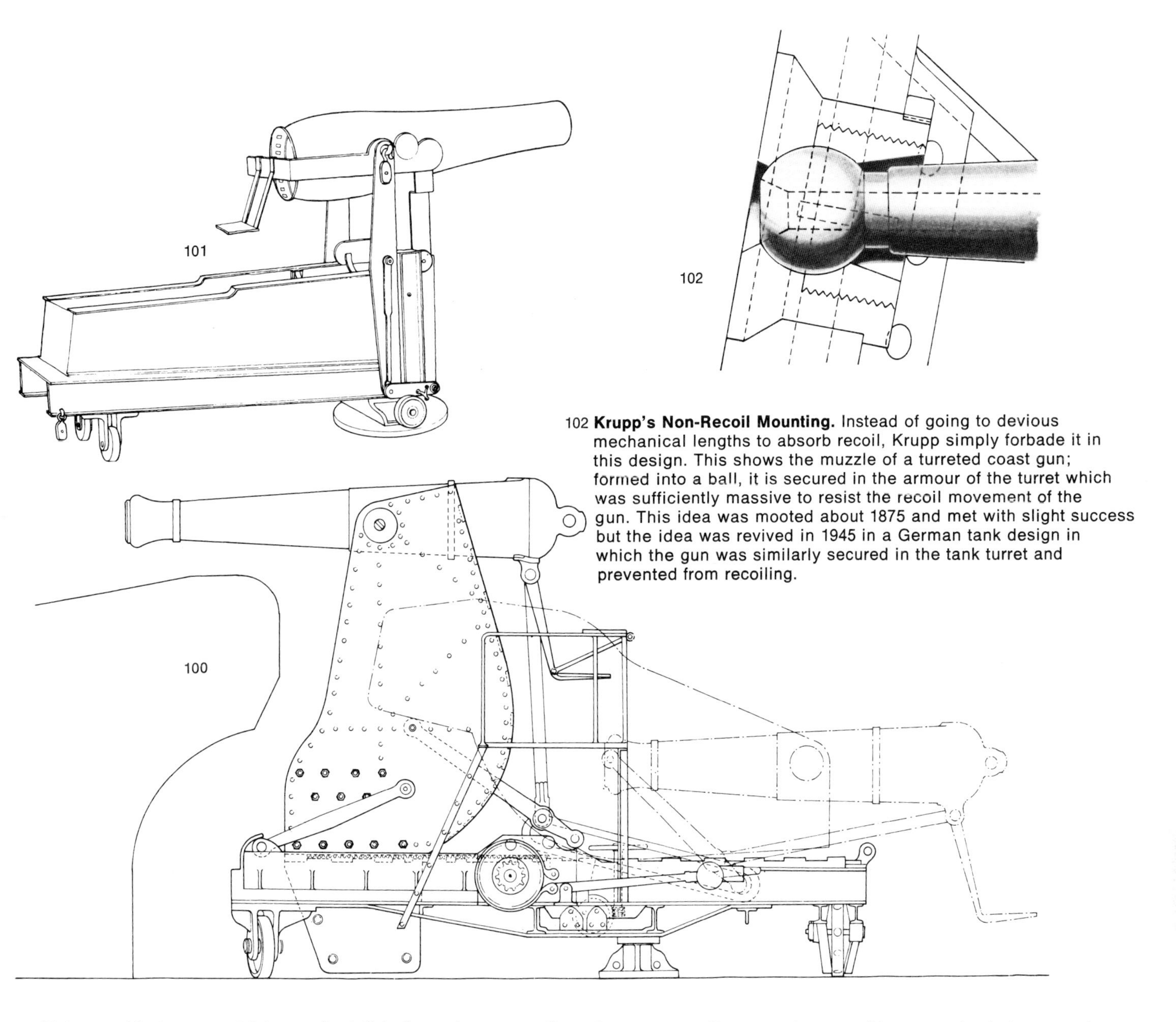

102 **Krupp's Non-Recoil Mounting.** Instead of going to devious mechanical lengths to absorb recoil, Krupp simply forbade it in this design. This shows the muzzle of a turreted coast gun; formed into a ball, it is secured in the armour of the turret which was sufficiently massive to resist the recoil movement of the gun. This idea was mooted about 1875 and met with slight success but the idea was revived in 1945 in a German tank design in which the gun was similarly secured in the tank turret and prevented from recoiling.

This provided a gun which was invisible from the sea until the time came to fire, and which protected the crew while loading and serving the piece. The lower carriage could be moved to give traverse, and an arc and screw on the top carriage controlled elevation. The only snag lay in the sights, which were still of the direct foresight and backsight type, which meant that the gunlayer had to leap up on the platform and expose himself to lay and fire the gun. This was a small drawback though, and the Disappearing Carriage (as this class came to be called) rapidly caught the fancy of every nation looking for a suitable gun mounting.

On the Continent, coast fortification was inextricably mixed with land fortification, since most of the Continental countries had land frontiers which were considered to be more important and vulnerable than their coastlines. Vast and expensive trials took place of armoured cupolas, armoured casemates, concrete emplacements, and the guns to fit inside. Krupp produced one noteworthy idea, the non-recoil gun, where a 15cm gun had the muzzle shaped externally into a ball, fitted into a socket forming part of the armoured face of the fort. The recoil was resisted by the ball and socket joint; needless to say this was a breech-loader, though there is no record of it having been officially adopted by anyone. It is interesting to look forward to 1944 and to see Krupp coming up with a variation on the same idea for mounting guns rigidly in tank turrets, using the mass of the tank to resist recoil.

Moncrieff, in 1877, produced an improved version of his mounting (100), which dispensed with the top carriage and slung the gun directly on top of the "elevators", as the curved arms were now known. Over eighty of these were taken into service and installed in British forts all over the world. One of the features of this mounting was a friction brake which screwed up tight to hold the gun in the loading position, being released to allow it to swing up to the firing position. In an old copy of the *Textbook for Carriages and Mountings* dated 1884, I have found a

103 **British 13.5 inch Disappearing Gun.** The largest British gun to be mounted on a disappearing carriage, three were emplaced in the 1890s but they proved to be too complicated and were withdrawn and scrapped in 1911.

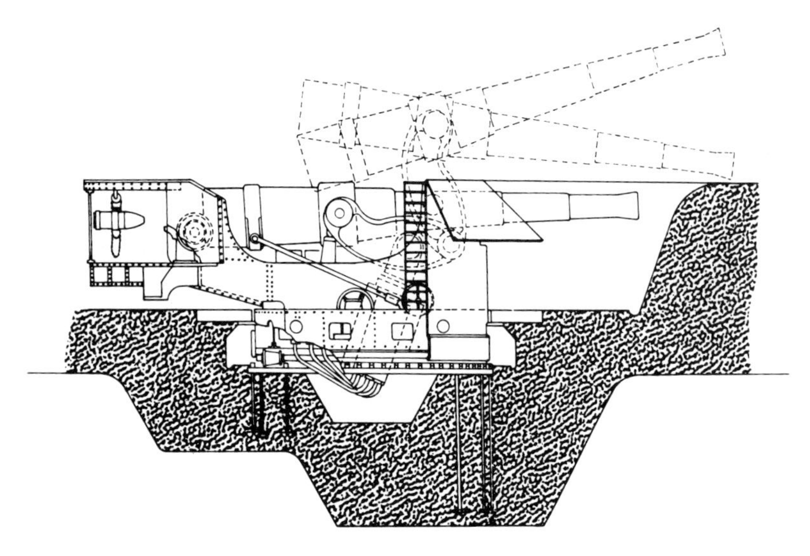

104 **British Hydro-Pneumatic Mounting.** The Elswick Ordnance Company married the hydro-pneumatic recoil buffer to the Moncrieff principle and produced this mounting which became the standard British pattern. Protected by a nine-foot deep pit and an overhead shield, they were emplaced in British defences from London River to Hong Kong.

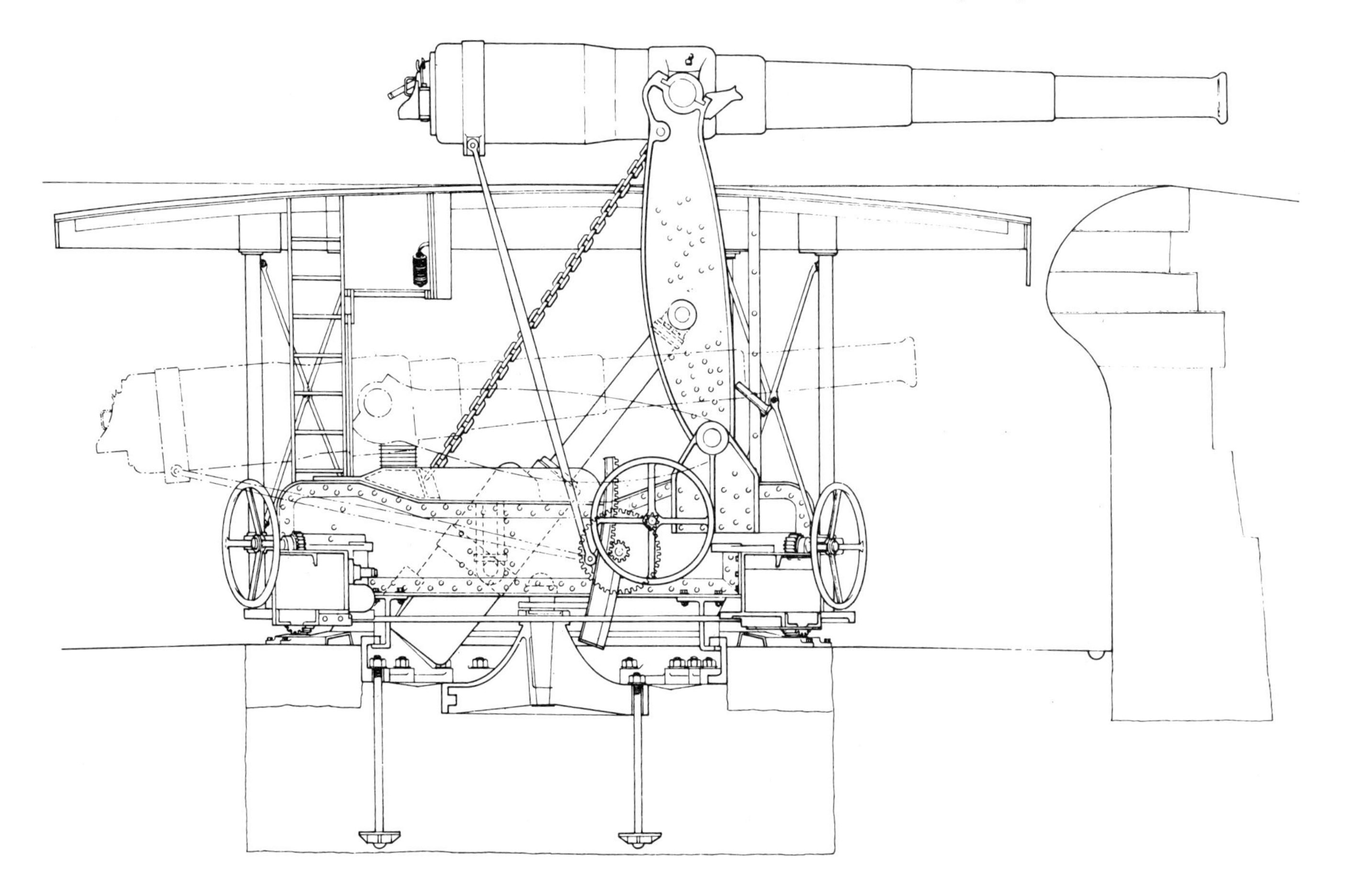

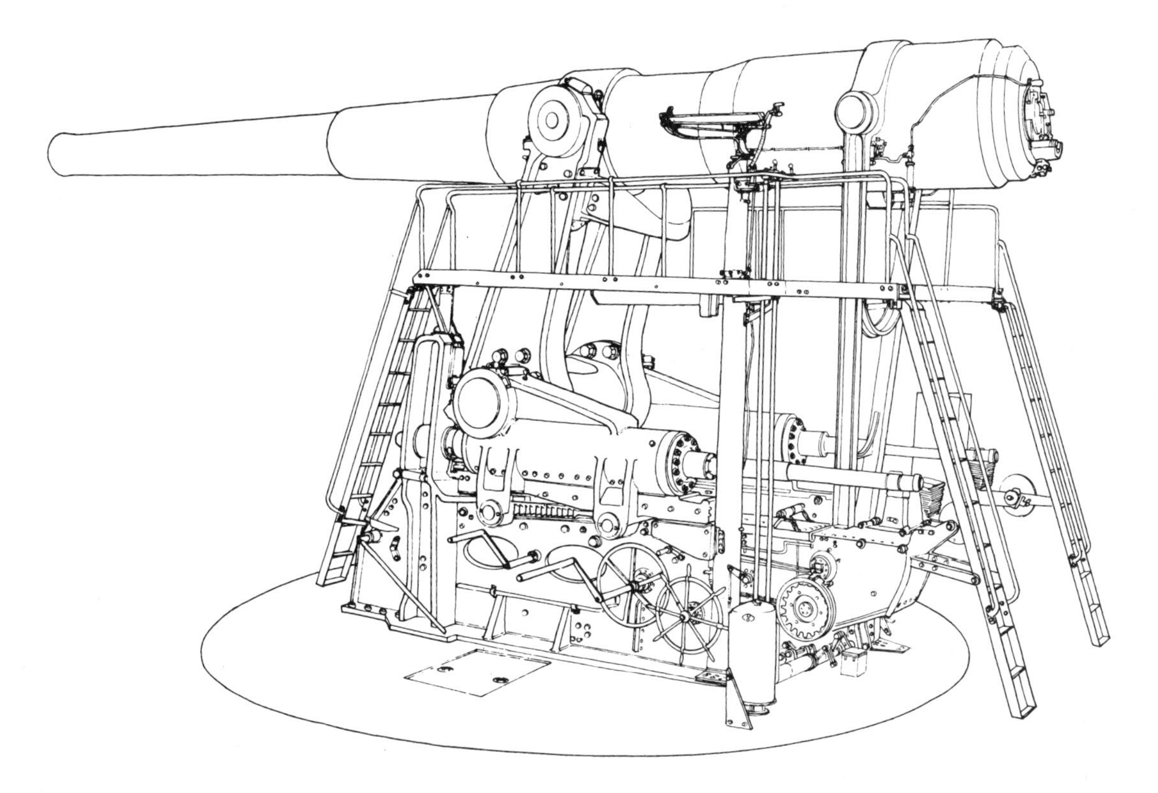

106 **The Buffington-Crozier Mounting.** The acme of disappearing carriage design, the American coast defences mounted everything up to 16 inch on these. Taking Buffington's idea, Crozier turned it into a practical device. The gun arms lift a massive counterweight and are damped by hydraulic buffers at the pivot point; due to the gun arm pivots moving back the gun describes a complex path during recoil.

105 A massive Krupp sliding carriage gun in Fort Chemenlik, Chanak, Turkey.

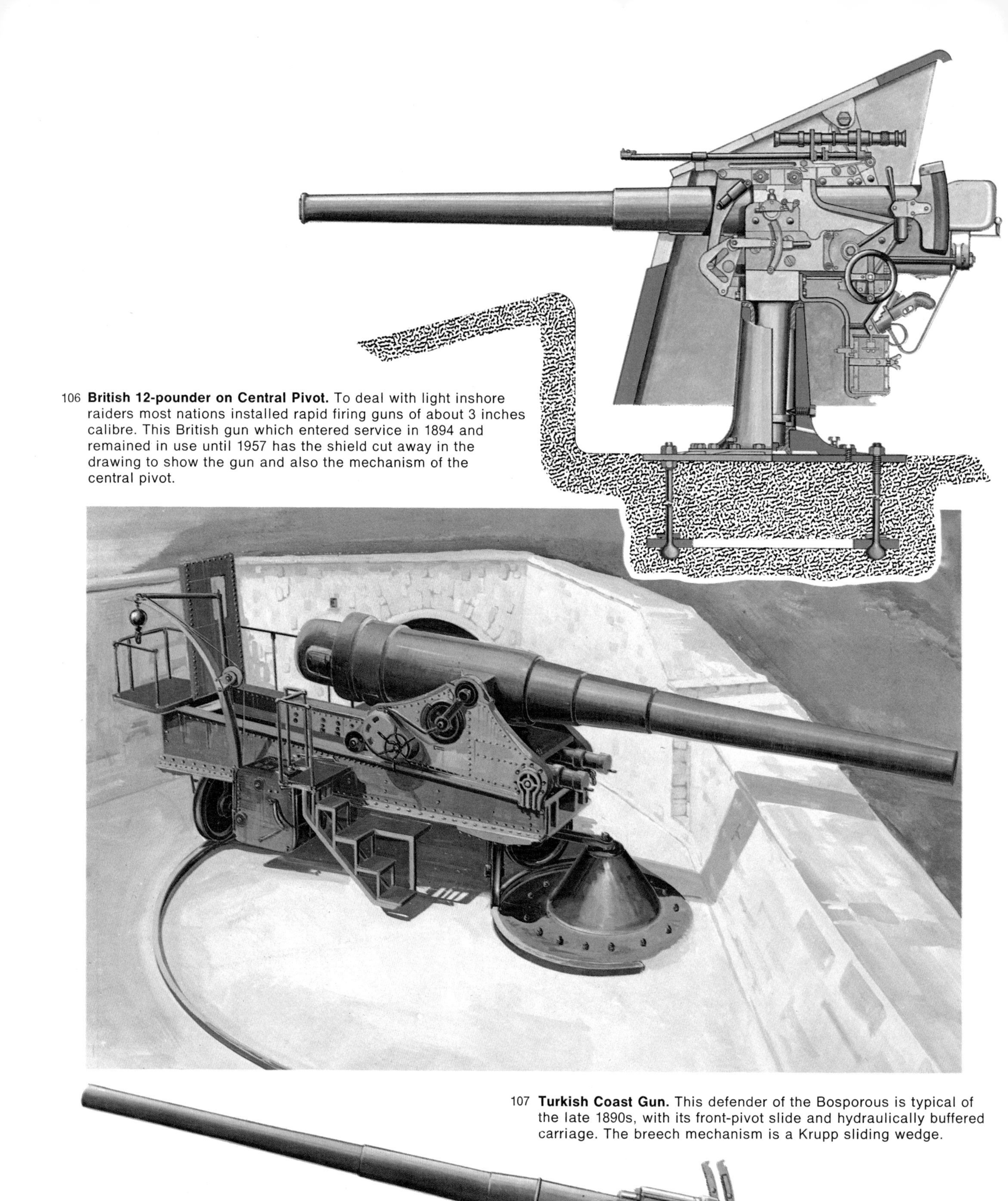

106 **British 12-pounder on Central Pivot.** To deal with light inshore raiders most nations installed rapid firing guns of about 3 inches calibre. This British gun which entered service in 1894 and remained in use until 1957 has the shield cut away in the drawing to show the gun and also the mechanism of the central pivot.

107 **Turkish Coast Gun.** This defender of the Bosporous is typical of the late 1890s, with its front-pivot slide and hydraulically buffered carriage. The breech mechanism is a Krupp sliding wedge.

108 **US 16 inch Rifle on Barbette Carriage.** The most powerful coast defence guns ever installed, these were used to defend such vital installations as the Panama Canal.

109 **US Coast Mortar.** In a similar fashion to the British, the American coast defences embodied large numbers of mortars for deck attack. Breech-loading, they were originally sited four to a pit in order to thoroughly cover a target, but as accuracy improved they were installed in pairs, which made service easier by thinning out the number of men dashing about the pit.

110 **The Depression Rangefinder.** By using simple trigonometry, coast gun rangefinding became easy. Here the rangefinder measures the depression angle to the target vessel. Since the height of the rangefinder is known, and since the angle at sea level must be 90 degrees, the range is easily calculated using schoolboy formulae. Obviously a correction has to be inserted to compensate for the changing level of the tide.

For a given height 'H', a relationship exists between range 'R' and depression angle 'D'. Thus with the sighting telescope of the DRF laid on the waterline of the target, the depression angle read off can be calibrated in range.

111 **Armour Piercing Shell.** The AP Shell may look simple but it took years of trial and research to perfect it. The tip must be hard to pierce the target; the body must be resilient to withstand side-stresses during penetration. The explosive must remain inert during the shock of arrival and the fuze must stay in place until penetration is completed and then perform efficiently at the proper time.

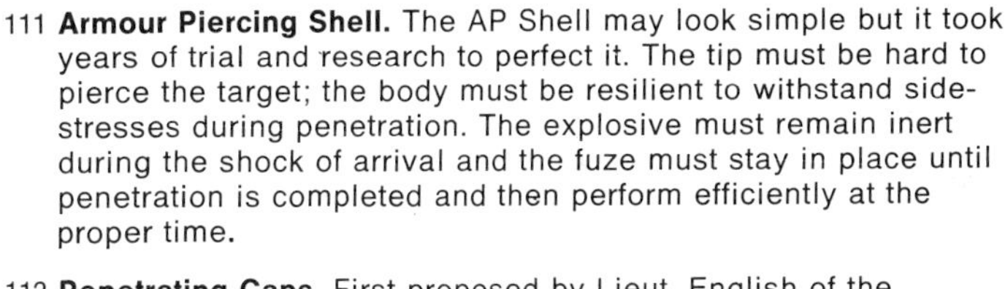

112 **Penetrating Caps.** First proposed by Lieut. English of the Royal Engineers and then forgotten; revived by Admiral Makaroff of the Tsar's Navy and adopted world-wide, the penetrating cap absorbs the shock of impact, spreading it on to the shoulders of the projectile, supports the tip during initial penetration, and to some extent actually 'lubricates' the passage of the projectile through the armour.

113 **The Mechanism of Penetration.** When the shot strikes the target if first deflects towards the plate, away from its line of flight (left). It then pierces the plate (centre) and then rotates to penetrate almost at right angles (right). It is this convoluted path which demands careful heat treatment of the shot body to prevent it shearing off.

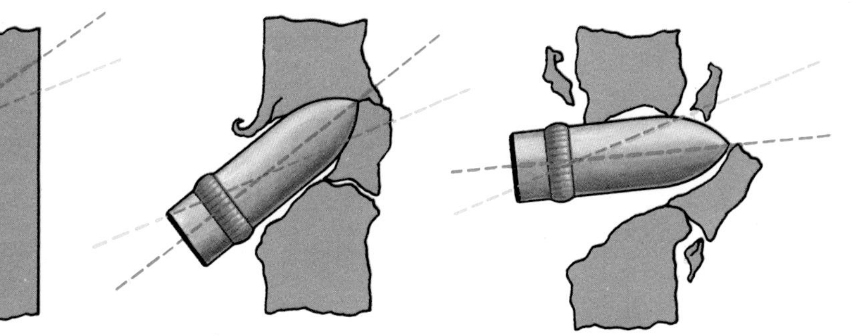

manuscript note added by some long-past instructor to warn his students of the consequences of absent-mindedly screwing the brake up once the gun had risen – apparently in 1881 a gun in Fort Camden, Cork, was fired with the brake applied and wrecked its mounting. The Moncrieff may have been an improvement on the sliding carriage but it was less soldier-proof.

When the hydraulic cylinder entered on the scene, the Elswick Ordnance Company, Sir William Armstrong's company, applied it to the disappearing carriage with conspicuous success. Their design was adopted in 1886 for the new six-inch breech-loader. Here the gun was supported on arms pivoted on a lower, traversing, carriage, with a hydro-pneumatic cylinder forming a strut at the rear (104). High-pressure air, acting on a hydraulic ram, kept the gun up in the firing position. It could then be hauled down by block and tackle for loading, which action further compressed the air. The air could be shut off from the ram by a valve to retain the gun "down" until it was ready for firing, then opening the valve allowed the air to push the liquid back into the ram cylinder and so force the gun up into the firing position, the valve automatically closing as the gun reached the correct point. Sighting could be done either directly from a platform or indirectly

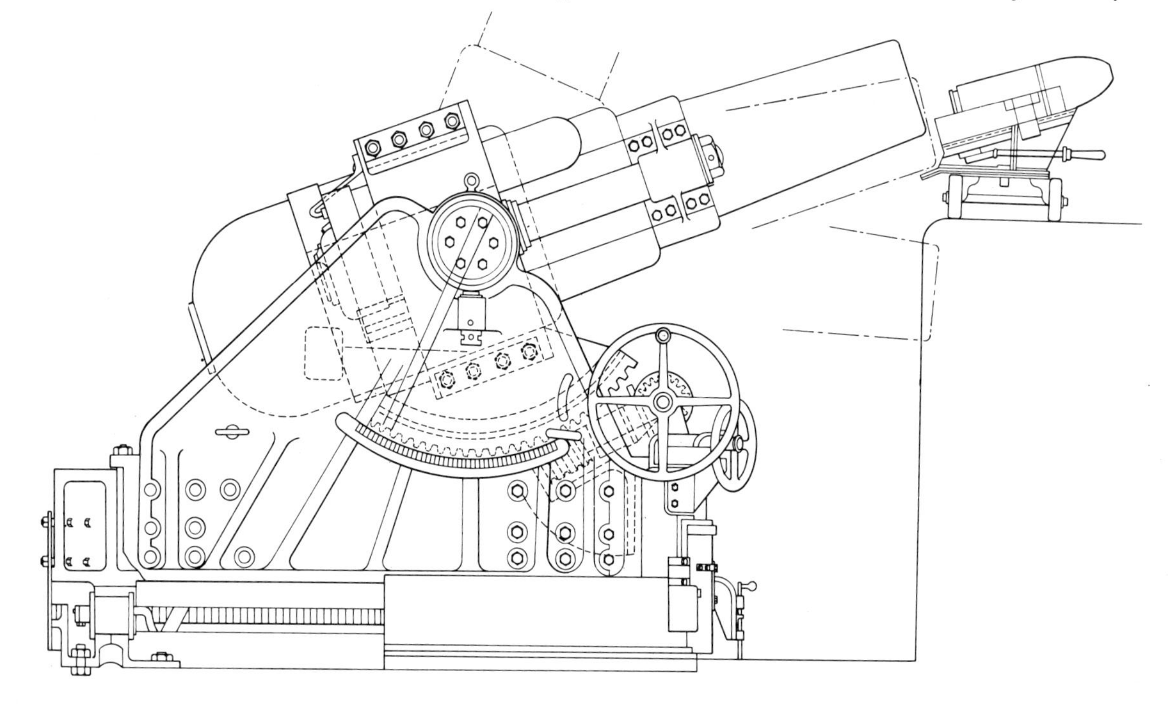

114 **British High Angle Mounting.** In order to attack the thin deck armour of warships, heavy guns were mounted so as to deliver plunging fire. These elderly 9 inch Rifled Muzzle Loaders found a new lease of life in this role.

115 **British Barbette Mounting.** The Barbette Carriage replaced the Disappearing pattern since it was simpler, allowed greater elevation, and gave a faster rate of fire. In this British 9.2 inch design the ammunition is handled in the pit, and the floor of the gun platform acts as an overhead shield. Hydraulic hoists feed the ammunition up to the gun.

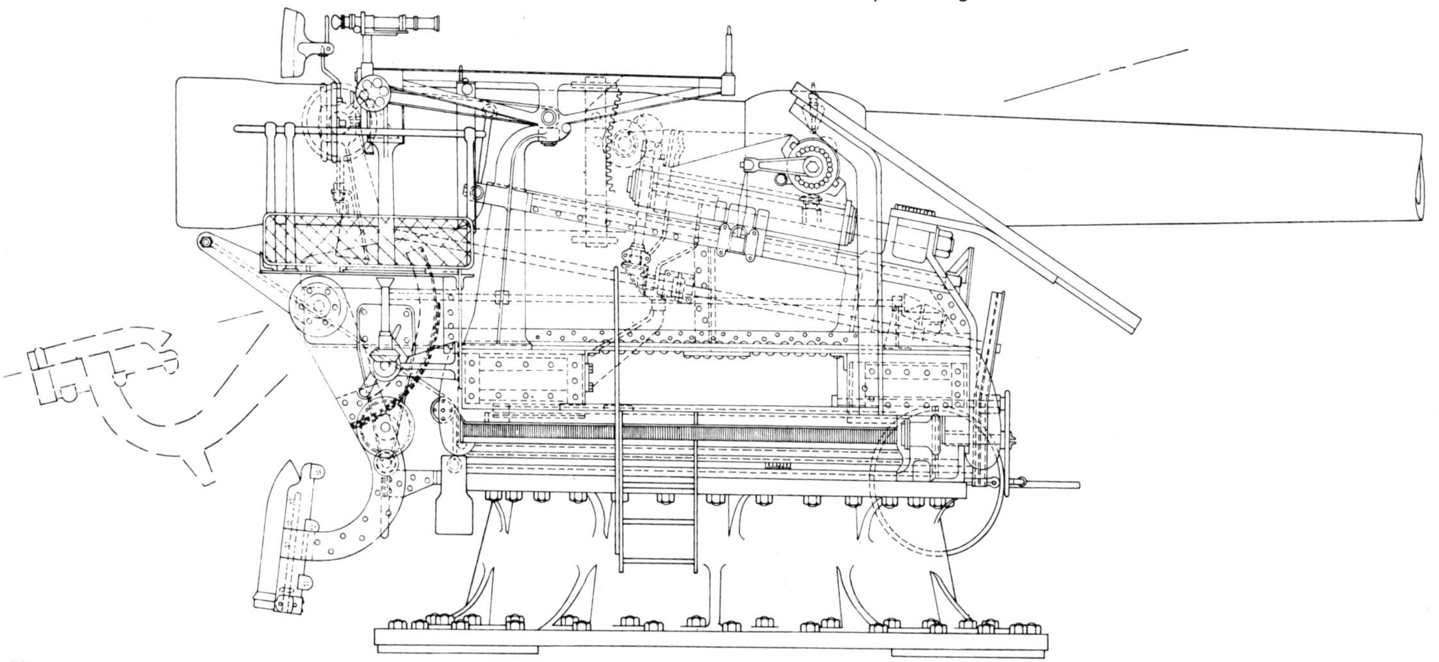

using a form of mirror sight which Moncrieff had developed with his improved carriage. When the gun fired the recoil forced the arms back and the moving piston displaced liquid back into the air tank through a spring-loaded valve so that the gun remained in the loading position until the valve was again opened.

In the USA Captain (later General) Crozier of the Ordnance Department had taken an early idea attributed to one General Buffington, in which the gun was mounted on a parallelogram, modified it and added hydraulic cylinders and a counterweight to produce the Buffington-Crozier Disappearing Carriage (101) which was the zenith of this type of carriage. Such was the geometry of this that instead of describing a simple arc, the gun moved back for some distance before swinging down, thus affording better protection to the men below by allowing the parapet to be that much closer. Guns of up to 14 inch calibre were commonly mounted on this carriage, and the last and biggest were the two 16 inch guns mounted at Panama during the First World War and dismantled late in the Second World War. (I am told that the only Buffington-Crozier left in existence is at West Point; I sincerely hope they treat it with the reverence it deserves.)

The disappearing carriage had two drawbacks which were insurmountable. One was the restricted elevation available – the greatest elevation ever managed by one of these designs was a mere 20 degrees and that was not good enough for the new and powerful guns which were coming into service at the end of the 1890s. The other drawback was the slow rate of fire due to the time taken for the gun to swing up and down and be re-loaded. In the early days of coast gunnery shooting was a leisurely affair, but the improvement in the speed of warships demanded an increased rate of fire and the disappearing carriage was at a disadvantage compared with a gun which stayed in position. Such a gun could be laid continuously on the target while loading and firing went on, another advantage. The demand for protection was now not so great, since the more powerful guns could keep the enemy ships out at a range where they could hardly see the shore weapons, particularly if some intelligence was applied to the matter of concealment and camouflage.

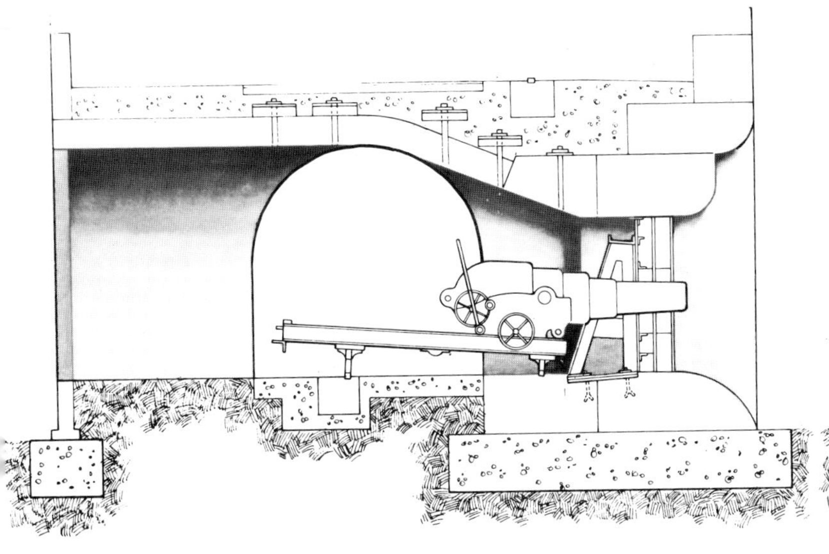

116 **Fortress Construction.** Examples of the massive concrete, granite and iron armour construction of nineteenth-century coast fortresses. At the left the armoured shield is supported by a frame, while the later version, on the right, seats the armour directly in the granite facing of the casemate, where it was secured by running molten zinc around it.

117 **The Maunsell Forts.** Located in the Thames Estuary and the Mersey, these Second World War forts were designed to mount anti-aircraft guns to close the water gap in the defences of London and Liverpool.

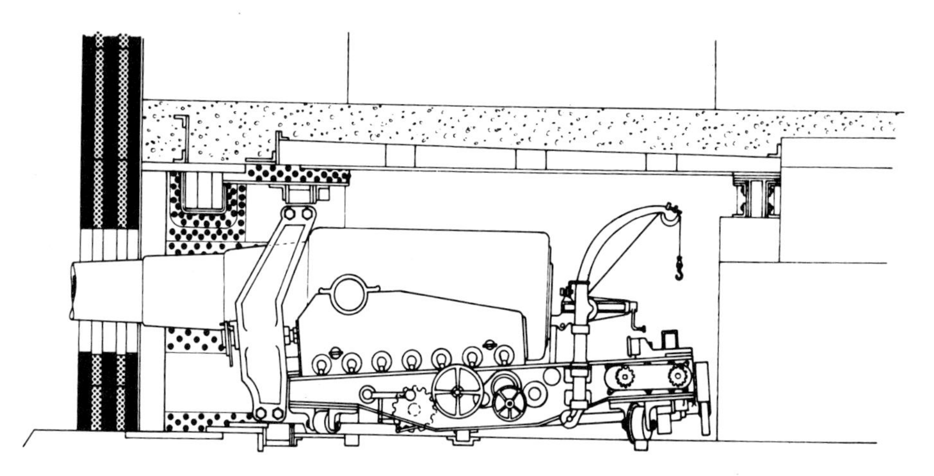

118 **The Spithead Forts.** The 12.5 inch breech-loading guns of the Spithead Forts in the Solent, showing how the yoke mounting was devised to spread the recoil force between the roof and the floor. This drawing also shows the construction of the iron forts, the slide and carriage and the loading davits.

Because of all these factors, by the late 1880s the barbette carriage had made its appearance, and from then on it remained the coast artillery's standard type of mounting (115).

The barbette carriage was generally emplaced behind a parapet or in a sunken position, behind a sloping glacis of concrete and protected by a shield forming part of the mounting. The carriage either pivoted on a solid steel pintle or rotated on a ring of rollers riding on a "racer" of steel set in the concrete, depending on the size and weight. The ammunition supply was protected by the parapet and fed up through lifts to the gun floor. The first barbette carriages used a modification of the old sliding carriage principle called the "Vavasseur" mount, named after its inventor, in which the gun recoiled up a short inclined slide controlled by hydraulic buffers, running back by gravity. This was soon superseded by a cradle type of mounting which permitted axial recoil. Since the mounting was heavy, and also securely anchored to the ground, the recoil and run-out were short and fast, and the service of the gun was considerably speeded up.

119 **The Twin Sixpounder.** What these British guns lacked in shell weight they made up in volume of fire. Two guns were mounted in the turret so that the gun captain could deflect them independently of the sights in order to compensate for wind, incorrect rangefinding or target movement.

In order to re-arm some of the older forts it was often necessary to go to complex lengths in designing special mountings to adapt modern weapons to old platforms. The older RML guns were generally placed in casemates, small chambers having loopholes through which the guns could fire. In order to keep the loopholes to the minimum size and thus present the smallest target to the enemy, the simple standing carriage was anchored to the front wall so that the traversing pivot ran in an imaginary line through the muzzle. This gave a small dimension width-wise but demanded a tall slot to cater for elevation. When naval vessels began to arm themselves with machine guns capable of shooting accurately through these openings, and guns began demanding more and more elevation, the dimensions were reduced without affecting the guns by adopting the "small-port" carriage in which a hydraulic jack was used to elevate the rear end of the gun, giving the elevation pivot also an imaginary line through the muzzle.

During Britain's 1860s building programme three forts were built in the Solent, between the Isle of Wight and the mainland, to protect Portsmouth Harbour. These, the Spithead Forts, were constructed of iron armour on concrete and stone foundations and they were originally armed with RML guns. In 1884 they were to be re-armed with 12 inch breech-loaders, and the original intention was to place them on Vavasseur mountings. But these would have to be shortened in order to fit them inside the casemates. Such a shortening would reduce the weight to the point where the guns would begin to jump, and the prospect of a 12 inch 43 ton gun plus twenty tons of mounting leaping about inside an iron fort in the middle of the sea was considered a trifle hazardous – and probably damned noisy as well. A special mounting was therefore devised which took the form of a heavy iron yoke through which the gun barrel recoiled. The modified Vavasseur carriage was behind, with the gun anchored to the yoke through two massive recoil cylinders (118). In this

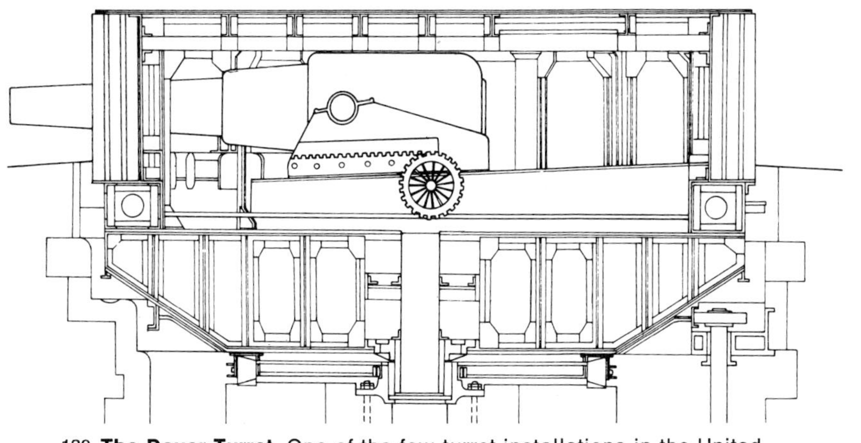

120 **The Dover Turret.** One of the few turret installations in the United Kingdom and the only one to mount muzzle-loading guns, the Dover Turret was installed in 1882 with two 16 inch 80 ton guns. They were rarely fired due to misgivings as to the state of the local cliffs. Thought declared obsolete in 1902 they were never dismantled, were recently rediscovered and are in process of being restored.

way the recoil stresses were evenly divided between floor and roof, and the dreaded jump was effectively damped out.

Since the primary target of coast fortifications was the ship, it followed that the armament of forts kept close step with that of ships. If a hypothetical enemy were to produce a ship with, say 12 inch guns, then one's own navy would have to follow suit and so of course would the coast forts who might be called upon to fight the same ship. Because of this there is a definite correlation in most countries between ship and coast armament at any time, and in most cases the ordnance was identical, only the mounting being changed according to the service. As we have seen, in the early days even the mountings were similar, and towards the 1910 period we find a school of thought which tended to bring the mountings even more into line. The basic idea was to lift the by now sophisticated ship's turret and fire control system and drop it straight into a concrete bed as a shore installation. This had several attractive features, such as commonality of components and an all-round fire capability. The idea had been flirted with in the Dover Turret, with its muzzle-loading 16 inch guns in the 1880s (120). It was revived in Britain during the First World War when two turrets to mount 12 inch guns were constructed to protect the mouth of the River Tyne. Germany took the armoured cupolas invented by Gruson, a Ruhr steelmaster who built up a considerable reputation as an armour specialist before being amalgamated with Krupp's; his turrets were flattened hemispheres of armour plate, set in concrete, and armed with one or two guns of calibres up to 24 cm, but these fell short of the complexity of naval construction. This was not on account of the technical problems, but simply that on land there was not the pressing demand to condense everything into one assembly, since space was not so limited as on shipboard. The full naval turret assembly was only of value where land was at a premium, and its greatest application came in just such circumstances.

The US naval bases in Manilla were defended by several fortresses, of which Corregidor was probably the most famous, but technically speaking the most impressive was Fort Drum on El Fraile Island. The water area south of Corregidor was too wide to be covered by guns on shore and on Corregidor, and in about 1908 it was decided to fortify El Fraile. The island was so small that the final solution was to encase it completely in concrete in the general shape of a battleship and install two turrets of naval pattern with 14 inch guns. From the air it looked like a stationary man-of-war and was promptly dubbed "The Concrete Battleship". In the 1942 siege of Corregidor its garrison remained in action right to the last, though parts of the fort had twenty feet of concrete removed by constant bombardment, a testimony to the construction and the robustness of the turret installations.

In step with the general improvement in armament there had been an equal improvement in fire control and tactics. In 1850 the defence of coasts had been a rather individual affair; the Fortress Commander gave the order to engage, and from then on it was a matter of every gun firing individually as the gun captain thought fit. The regulations in force at that time in Britain simply said, "The serjeant will place himself to windward of the gun and observe the strike of his shot so that he may make such corrections as he feels are necessary." And that was Fire Control in 1850. This was gradually improved, first by grouping guns into small fire units under a commander so that concentrated fire could be brought to bear on a selected target. Eventually whole forts were welded into coherent fighting units, but this had to wait until suitable instruments were developed to aid the gunner. The greatest problem was the the determination of range of a moving vessel. Captain Watkin, a garrison artilleryman who made an intensive study of rangefinding and similar problems, developed his first "Depression Range Finder" in 1873, an instrument which relied on the trignometric relationship between an instrument at a known height and a distant target producing a specific depression angle for any given range (110). The DRF, as it became known, at least gave the range accurately and soon showed an improvement in shooting, but with the increasing speed of vessels and the increasing ranges of engagement, the problem now arose of trying to forecast where the enemy vessel would be by the time the projectile got out there – a sort of two-dimensional fore-runner of the anti-aircraft problem. After more years of experiment and trial, Watkin produced his "position finder", which took into account tide, current, target speed and course and ballistic corrections, and produced a forecast position. The PF operator, who became a mysterious, highly-paid, scientific hermit, lived in an armoured cell away from the Fort and controlled the fire of the guns directly having communication with the gun-floors. Although this device took some time to overcome initial troubles (and the reluctance of commanders to hand over command of their guns to the hermit in his cell), once its results were seen it rapidly gained favour. One by-product of the PF was a vast improvement of the drill at the guns, since the gun captain could no longer take his time over re-loading and stroll off to windward for a casual appraisal of the situation while his detachment laboured. Now he had little more to do than crack the whip and see the gun was loaded as fast as possible and laid in accordance with data supplied from the PF; it was no use having a device which predicted the target's position if the gun wasn't ready to fire when the PF demanded it. Consequently the rapidity of serving the guns increased, and the coast artillery gun drill began its evolution into a drill so slick as to resemble an intricate ballet.

So far we have largely considered the heavier guns since these represented the first line of defence and were the weapons originally installed to counter the armoured vessel. But in the 1880s the motor torpedo boat began to show its face on the water and naval manoeuvres of the 1890s showed that most nations were considering the use of these fast and small vessels for lightning raids on enemy docks and harbours. In order to counter this threat small fast-firing guns in large numbers were wanted. In 1880 Hotchkiss and Nordenfelt had produced their quick-firing 3pr and 6pr guns which had been adopted by many navies, and in the mid-1880s these began to be installed in great numbers both in Britain and elsewhere, for defence against torpedo boats. The 3pr was a little light for the task, and before long was relegated to a practice role, while the British defences were strengthened by adopting a 3 inch 12 pounder (106). The deployment of these guns were backed up by searchlights, used to illuminate areas of sea and termed "Fighting Lights", and the impressive volume of fire which these light guns could deliver boded ill for the "mosquito boats" – a current description.

So far the method of attack had been simple. See a ship and shoot at it. The point of aim was the side of the vessel, and you hoped to slam an armour-piercing shell through to burst inside. To counter this, of course, the naval architects made the ship armour thicker. So the gunners of the next generation got themselves heavier guns. And so it went on. But in the 1880s another inspiration arose, both in Britain and the USA. While the armour of the sides was getting thicker, the armour of the deck remained more or less the same – relatively thin. Moreover making the deck armour the same thickness as the side would render the ship liable to roll over, so there was a practical limit to deck thickness. Let us therefore drop shells down onto the deck instead of letting ourselves get caught up in a race for heavier guns.

Britain started in a typically cautious fashion by re-lining a spare RML 9 inch gun, building a howitzer-like mounting, and performing a few trials. The idea worked better than anyone had hoped, the trial gun securing direct hits on a moving target at ten thousand yards range and demolishing deck armour with the greatest of ease. As a result, a number of RMLs were re-rifled with polygroove rifling, mounted on these "high angle" mountings, and installed at various ports (114). One of their great advantages of course was that they had no need to be in view of the enemy; they could be hidden away in any convenient fold in the ground. Later a breech-loading 9.2 inch was also fitted to this type of mounting and a number were installed at Plymouth. But Britain never took to the high-angle gun as the Americans did; their versions too came out in the 1880s and their 12 inch mortars were mounted in large pits, four weapons per pit, in great numbers. Later the confusion and uproar in the pit when the re-loading of four guns was in progress led them to reconsider and move to two-gun pits, but these weapons were a highly-regarded component of US coast defence (109). They were even used for land bombardment of Bataan from Corregidor in 1942.

The First World War found the coast artilleries of the world at the top of their form. All the technical and tactical developments of the 1890's had been assimilated, armament was generally as good as could be afforded, and everyone sat down to await the onslaught. And, by and large, they went on waiting. For very little coast artillery was engaged during that war; the Germans bombarded Hartlepool in England and provided lively employment for a brace of six-inch guns. The Allied Fleets sailed up to the Dardanelles and took a bloody nose for their pains, bearing out Admiral Fisher's dictum "Nobody but a fool attacks a fortress". And that was about all. There was additional fortifying of various places as the tides of war ebbed and flowed, but no decisive coast artillery engagement.

When the war was over, there was a massive thinning-out of coast defences all over the world. This was partly due to the general disarmament mood and partly to revised tactical thinking. Many of the older works with casemates and disappearing carriages dated from the days when guns had so little range and such slow rates of fire that it was necessary to over-insure to make sure there would be enough shells flying through the air to stop the enemy. With modern faster-firing weapons and their longer ranges, less guns could cover the same areas. Consequently both Britain and the USA had bargain sales of redundant fortresses in the 1920s and re-worked their remaining defences to use modern mountings – so far as the national purse-strings allowed them.

The principal strengthening of coast defences during the inter-war years was done by Britain in Singapore, and, of course, by Germany during her re-armament. Singapore saw the installation of five 15 inch guns, the most powerful armament ever installed by Britain. The German installations were relatively sparse and were of all calibres up to 38cm, largely naval guns, since the coast defence of Germany was in the hands of the Navy, probably on the theory that poachers make the best gamekeepers.

The US Army placed a good deal of faith in railroad equipments for rapid deployment, as we have already seen. Another system was to prepare concrete emplacements at various places onto which the standard 155mm M1917 towed gun could be rapidly fixed in time of need. This technique began in the Canal Zone, and the general type of mount soon became known as the "Panama Mount" irrespective of where it was installed.

The Second World War brought the coast gunners the same pattern of long periods of boredom interspersed with bouts of furious activity. There were more actions in this war, simply because the active operations spread all over the world. The earliest were the actions by the Norwegian forts defending Oslo, followed by the fall of Singapore – though this was not a coast artillery action – and the fall of the Manila defences – although again, this was not strictly a coast artillery action. Malta in 1942 did a lot to restore faith in coast armament, when Italian torpedo boats attempted to attack the Grand Harbour and were blown out of the water by the British twin-barrelled 6-pounder (119), the modern version of the original 6-pounder anti-torpedo-boat gun. Firing at 120 rounds a minute it was a remarkable torpedo boat which survived an encounter with the twin-six.

Towards the end of the war the guns mounted in the Dover area were able to get in some practice against their rivals, the German 40.6cm guns on Cap Gris-Nez. These two groups had shelled each other off and on since 1940, but such was the range and difficulty of observation that it was more in the nature of a morale-builder (or destroyer, depending upon which end you were) than serious gunnery. Once the Allied armies had landed on French soil, an air observation post Auster airplane was allotted to the Dover guns and they opened fire on Cap Gris-Nez for the last time. Within three rounds they had a direct hit, and in conjunction with a land attack in the area the German guns were silenced. It must rank as the longest-range infantry support shoot ever fired.

The Second World War was the last stand of the Coast Artillery. The US disbanded their units shortly after the war, and the British followed suit in 1956, the annual "Statement on Defence" saying, "In the light of modern weapon developments it is clear that there is no longer any justification for maintaining Coast Artillery. The seaborne threat can be countered more effectively by the Navy and RAF and other types of artillery can, if needed, be used for seaward defence." It was the end of an era.

121 **Krupp Anti-aircraft Gun.** This Krupp design was exhibited in 1909 and was copied elsewhere in subsequent years. With the point of the trail anchored, and the wheels swivelled round, the whole weapon could be traversed very rapidly in order to keep up with a moving airplane.

122 **Krupp Balloon Gun.** Another 1909 exhibit at the Frankfurt Fair, this was a 12-pounder intended for anchoring in concrete as a static defence weapon.

123 **Krupp-Daimler Motor Gun.** An early contender in the self-propulsion field was this 1909 motorised gun, an example of the early doctrine of pursuit.

124 **The One-Pounder Pom Pom.** In the early days of vertical gunnery, with rudimentary sights and primitive fire control, it seemed that a high rate of fire might make up for a lot of the deficiencies. As a result of this theory the pom-pom was placed on high angle mountings. But once aeroplanes learned to stay high enough, its day was over.

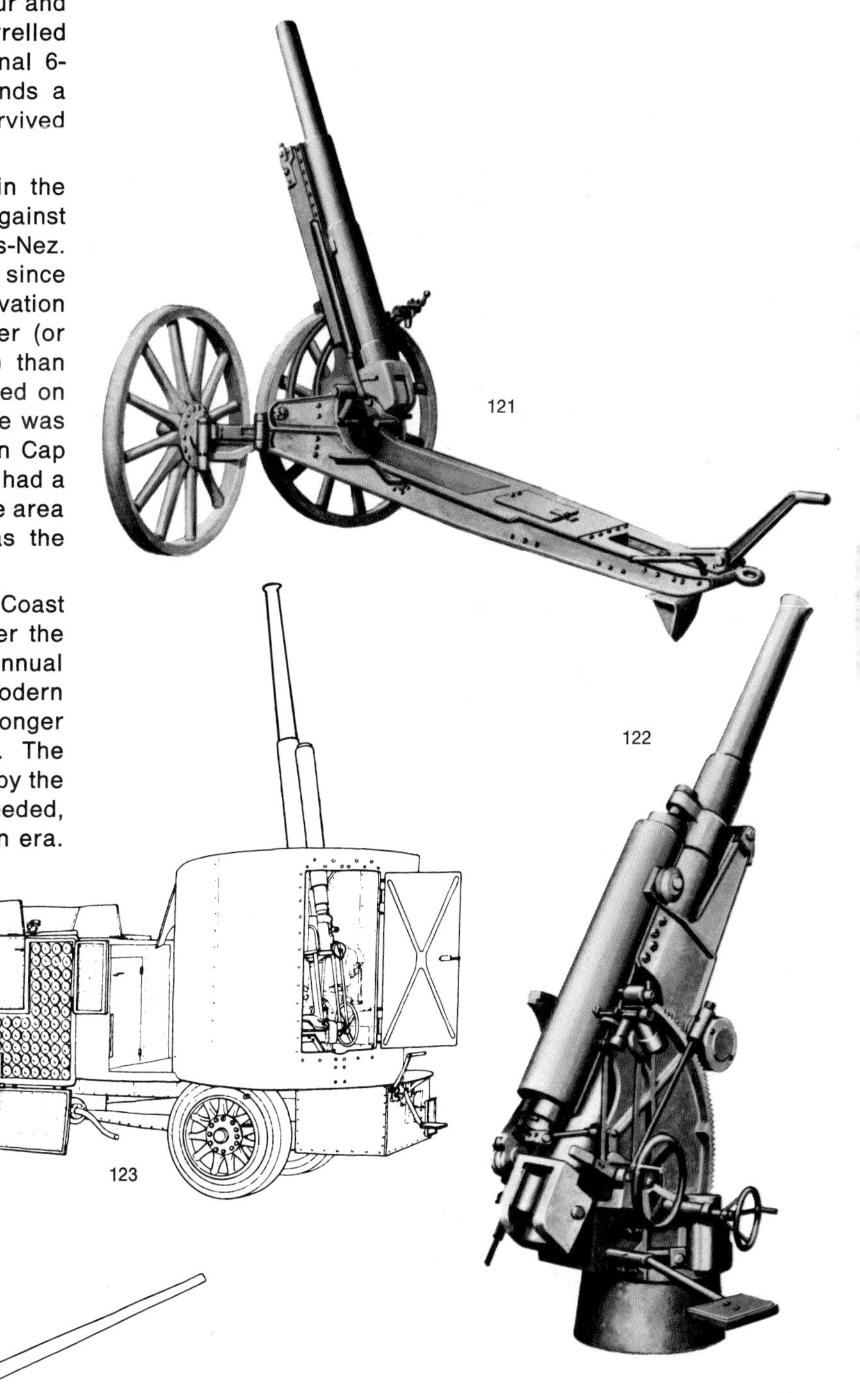

125 **German 7.7cm AA Gun.** An attempt to turn field guns into anti-aircraft guns with the least modification is this German method of mounting at high angle on a rotatable staging. While it is a workable idea, the balance of the recoil system is severely upset, having to return the gun into battery at such a steep angle.

126 **French Autocannon.** The universal French response to any artillery problem was the 75mm Gun, and so their first anti-aircraft weapon was a Model 1897 on a De Dion chassis. It was, in fact, remarkably successful in this role and numbers were later supplied to various allies.

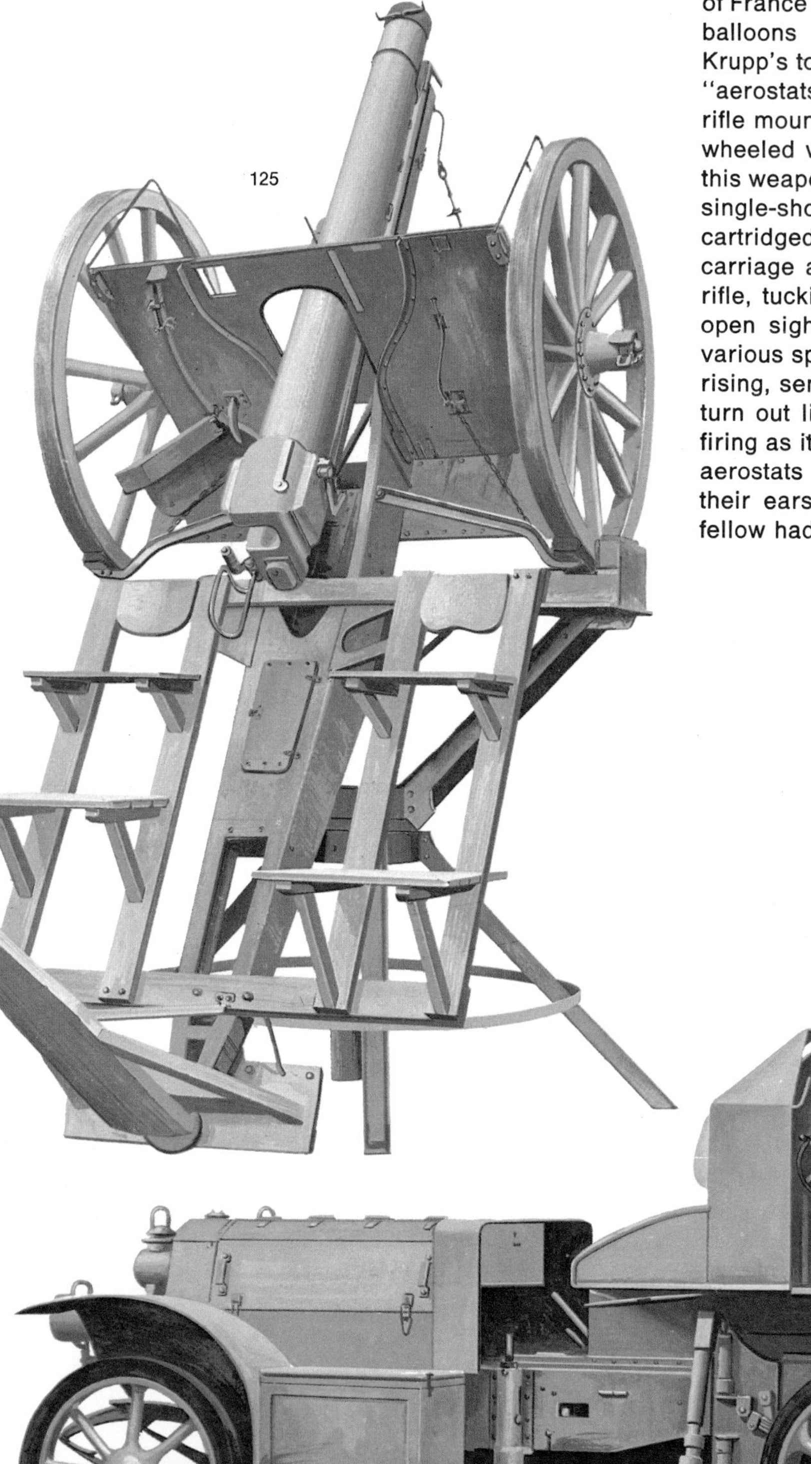

Anti-aircraft guns

No sooner had man succeeded in conquering the air than soldiers began to contemplate the prospect of knocking him back down again. The first serious attempt at anti-aircraft shooting was made outside Paris in 1870; the Prussian Army had besieged the city, but some of its more ingenious citizens proposed to use balloons to fly over the investing troops and operate a mail service to that portion of France remaining in French hands. After a few of these balloons had successfully escaped, Moltke ordered Krupp's to manufacture a suitable weapon to shoot at the "aerostats". Krupp rapidly produced a one-inch calibre rifle mounted on a pedestal set on the bed of a light four-wheeled wagon pulled by two horses. Precise details of this weapon are scant, but it would appear to have been a single-shot weapon, using some form of breech-loading cartridged round. The gunner rode on a seat on the carriage and, when a target appeared, knelt behind the rifle, tucking the stock into his shoulder and firing over open sights. A number of these guns were located at various spots around Paris and when a balloon was seen rising, sentries would alert the nearest outfit. This would turn out like a fire engine and gallop after the balloon, firing as it went. Primitive as it may sound, many intrepid aerostats complained about the bullets whistling past their ears at heights of 800 to 1000 metres; one poor fellow had so many holes punched in his balloon that he was forced down in the Prussian lines and taken prisoner.

From then on the balloon, and later the aeroplane and airship, engaged the gunners' attention. The aeronauts, for their part, were more interested in finding out how high they had to fly to avoid reprisals from the ground. Some early aviators who acted as observers in the Balkan War had many close calls from rifle-fire, one or two unfortunates being killed, but these were considered to be unlucky exceptions. To do any real damage to the flying machines, something heavier than rifle-fire was going to be needed. The Frankfurt International Exhibition of 1909 saw the Continental gunmakers drawing back the curtains from around their drawing offices and displaying their answers to the new threat.

Krupp exhibited no less than three different guns, a 65mm 9-pounder, a 75mm 12-pounder and a 105mm. The 65mm had a clever two-wheeled carriage on which the wheels could be folded round to the front and the trail anchored so that the carriage could trundle round to give 360 degrees of traverse with an elevation of 75 degrees and a ceiling of 5700 metres (122). The 75mm was on a motor-truck mounting with a ceiling of 6500 metres, while the 105mm was pedestal-mounted for use in ships and claimed a ceiling of 11,500 metres; one cannot help feeling that the ceiling figures quoted were more in the nature of pious hopes than proven performances.

Erhardt went one better than Krupp and produced a 5cm quick-firer mounted in a turret on an armoured car. This display gave everyone something to think about. The following year saw the use of aircraft on military manoeuvres become a commonplace event, and plans for anti-aircraft weapons began to crystallise. France decided on

127 **British 13-Pounder on Peerless Lorry.** Britain tried the 18-pounder field gun as an AA weapon but it was not a success until it was modified by lining the barrel down to 3 inch calibre and using the 18pr cartridge to propel the 13pr shell. On this truck mounting it became an efficient and popular weapon, being retained in many other armies for some years after the war ended.

128 **German 8cm AA Gun.** Krupp's design of towed AA gun which was the final German gun of the First World War to enter service and which was the godfather of the World War Two 'Eighty-Eight'.

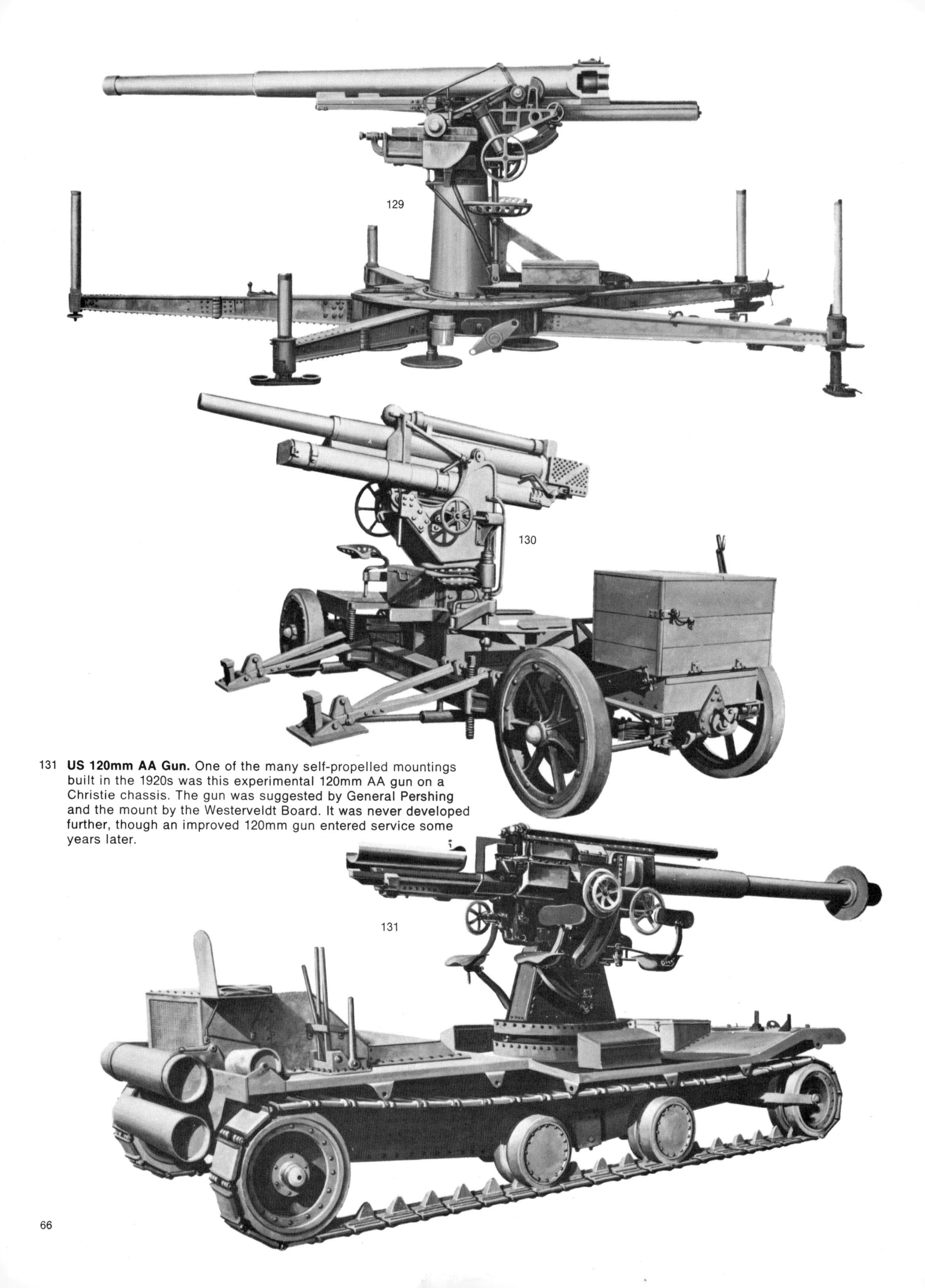

131 **US 120mm AA Gun.** One of the many self-propelled mountings built in the 1920s was this experimental 120mm AA gun on a Christie chassis. The gun was suggested by General Pershing and the mount by the Westerveldt Board. It was never developed further, though an improved 120mm gun entered service some years later.

132

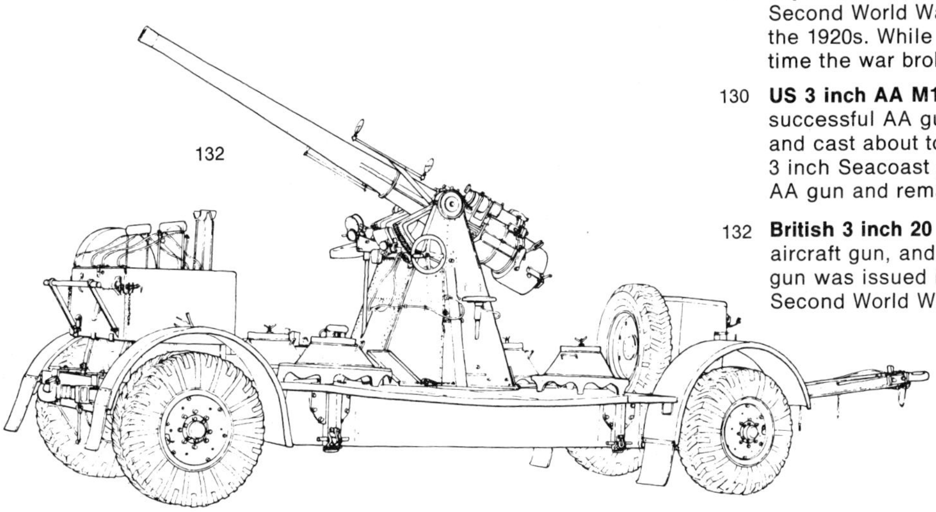

129 **Japanese 75mm Type 88.** The standard Japanese AA gun of the Second World War was based on a Vickers design purchased in the 1920s. While efficient in its day, it was obsolescent by the time the war broke out.

130 **US 3 inch AA M1917.** The US observed that most of the successful AA guns in Europe were of about 3 inches calibre and cast about to see what they could use. The answer was the 3 inch Seacoast gun, which on a new mount became the 3 inch AA gun and remained in service until the late 1930s.

132 **British 3 inch 20 cwt Gun** The first British 'purpose-built' anti-aircraft gun, and the innovator of the 'cruciform' mounting, this gun was issued in 1914 and remained in use well into the Second World War.

her standard 75mm M1897 mounted on a motor truck, while Britain designed a completely new 3 inch gun which, to distinguish it from others of the same calibre, became known as the 3 inch 20 cwt from the weight of the piece.

The outbreak of war in 1914 found the major powers with little anti-aircraft equipment available. The German Army had some Krupp 75mm and 80mm guns, Britain a few of the 3 inch 20 cwt, and France but two Autocannon. In order to fill the gaps, Britain called up one-pounder pom-pom guns and fitted some on pedestal mounts for defending fixed locations and some on high-angle field carriages for mobile use (124). France having a sufficiency of 75s – or so it thought – threw them onto trucks and trailers of every sort (126). To improve the heavier defences, Britain took the 13-pounder Horse Artillery gun and placed them on trucks – for there was still some element of the chase about anti-aircraft tactics in those far-off days (127). Eventually, practically everything that could be persuaded to shoot upwards was proposed or tried as an anti-aircraft weapon, including such remarkable ordnance as six-inch howitzers and three-inch trench mortars. When the initial panic had died down, the 3 inch, 13-pounder, 75mm, and, on the German side, 75mm and 80mm emerged as the accepted standard weapons, although many odd weapons held their place until the end of the war.

Having weapons was one thing, using them was a horse of a different colour. The first system of fire control was what an American called the "Guess-point-shoot-and-pray" system, and that fairly sums it up. For it was soon seen that here was a problem quite unlike anything which had been met before. Here was a target free in three dimensions, small, manoeuvrable and fast; in the time between setting the fuze, loading the gun, firing, and eventually having the shell arrive at the same height as the airplane, the target had moved a considerable distance, and it might have moved in any – or many – of several different directions (153). It became necessary to apply some scientific thought to this matter of shooting at airplanes, and in every country eminent scientists were called in to help.

While this was going on, another controversy was raging; what was the best projectile to shoot against these new targets? The odds on getting a direct hit appeared slender, therefore it would seem advisable to employ some shell with an area effect and to use a time fuze to burst it somewhere in the target area, hoping that the burst would envelop the airplane. Shrapnel was favourite, followed closely by High Explosive, but the outsider Incendiary was coming up strongly, and eventually proved, if not the winner, at least a close second to shrapnel. The virtue of shrapnel lay in the multitude of high-velocity bullets which could be flung out, any one of which was damaging to the aircraft and lethal to the occupants. High explosive at that time was held to be too chancy a proposition, for efficient detonation could not be guaranteed. The Incendiary, or "Anti-Zeppelin" as it became known, was similar to shrapnel but threw out pellets of flaming thermite, which could penetrate and ignite an airship or aircraft petrol tank.

While the technical problems were being faced, the tactical problems were giving the infant "Archie" a hard time. (Early AA was christened "Archie" or "Archibald". This term remained in use until the Second World War when, probably due to the influence of an early BBC radio show "Ack-Ack, Beer-Beer" – a morale-booster for personnel of anti-aircraft and Balloon Barrage sites – the term "Ack-Ack" came into use and ousted "Archie" from the field.) The most prominent feature was the shortness of engagement time, which became a vital factor in fire control. It was no use taking time over abstruse calculations, since the target would have passed out of range. So the general practice was to have a fire unit of but two guns, and have as much of the fire control system on the gun as possible. This, in turn, meant surrounding the gun with a team of highly-skilled operators busily setting sights, operating fuze-setting machines, calculating aim-off and fuze length, as well as the usual strong-arm gang who were doing nothing but point and shoot. While this system was reasonably efficient, it demanded a huge supply of skilled technicians and it was not long before the manpower shortage forced a rethink. It was a better and more economic proposition, it was felt, if one central post were to use all the technicians in producing data to pass to the guns, and the guns simply did as they were told and fired as fast as they could; the rate of fire would probably be improved when all the mathematicians were cleared away from the gun anyway.

The French made a considerable step forward, which was immediately taken up by Britain, with their adoption of the "Brocq Central Post Tachymeter", a device consisting essentially of a telescope mounted on a tracking head. When the assumed range was set into the instrument and the handle turned to track the target, a tachymeter – or speedometer – indicated the target speed, and, via charts and graphs, this could be translated into an accurate aim-off for the future position of the target. The normal range-finder was pressed into service, slightly modified, as a height-finder, and this enabled a fuze length to be calculated. But all this information and data referred to an imaginary gun occupying the position actually taken up by the Central Post Tachymeter, and the data had to be converted to suit the actual physical location of the guns, which might be displaced several hundreds of yards away. Eventually, having arrived at a convincing set of results, the data would be shouted or telephoned to the guns and fire would be opened. Once this system got under way, larger groups of guns became common.

By the end of the First World War enough had been done to show the direction in which anti-aircraft gunnery had to go in order to become successful, though for a time after the war there was a period when some voices opined that anti-aircraft gunnery had been nothing but a panicky attempt at defence, and that it could now be thrown away and reliance for defence placed on the design of fighter aircraft. Fortunately, saner counsels prevailed, and anti-aircraft gunnery began a long period of trial and experiment. The Central post principle was accepted as the only workable method of controlling fire; a means of sending the information to the guns in a foolproof fashion was needed, and above all some mechanical method of solving the fire control problem had to be found. The guns, too, had to be improved to keep up with the general improvement in the performance of aircraft.

During the 1920s Vickers of England threw themselves whole-heartedly into these problems. They developed a useful 75mm gun which was adopted by many countries, and also set to work on a "predictor" which would mechanically solve such problems as future height, position, range and fuze length. To go with this, electrical systems of data transmission were developed so that movement of the central instrument could instantly be passed to dials on the guns; the gunlayers merely laid their guns to the data on the dials and had no further need of optical sights. Vickers eventually produced their predictor which was adopted by several countries, though not immediately in Britain. The US Army began developing 3 inch, 105mm and 4·7 inch guns of high performance and a Major W. P. Wilson developed a predictor. The Sperry Gyroscope Company were asked to produce a data transmission system for this Wilson predictor and in doing this they managed to point out a few improvements which could be made to the Wilson device. The Army adopted both the Wilson and Vickers predictors, in order to find out which was approaching the problem from the best angle, and eventually decided that while the Vickers was weak in its computation it was very good at the prediction side of the problem, while the Wilson was a poor predictor but a good mathematician.

During the First World War there had only been one kind of anti-aircraft fire and it encompassed everything that flew. In the early 1930s though, as Air Force techniques improved, it was soon apparent that there were two separate and distinct types of target. There was the high-flying bomber, relatively slow and less manoeuvrable. And there was now the fast-flying low-level attacker, the ground-strafing fighter, the dive-bomber, planes which moved fast, flew low, and were highly manoeuvrable.

The first type of target was little different from the wartime pattern, though now flying higher and slightly faster, and the problems to be solved here were appreciated and – more or less – under control. The low-level attacker posed a new problem; the low altitude and high speed meant that the guns had to be capable of swinging and training exceptionally quickly in order to keep up with the fleeting target, and the time of engagement was very short.

133 **German 12.8cm Flakzwilling.** One way to put more metal in the sky is to produce twin guns; these German 12.8cm weapons were usually installed on top of Flak Towers in industrial areas, and proved highly effective.

134 **German 15cm Flak.** As with every class of gun, anti-aircraft weapons became bigger and bigger as the war went on. But this model proved too cumbersome and was never put into production.

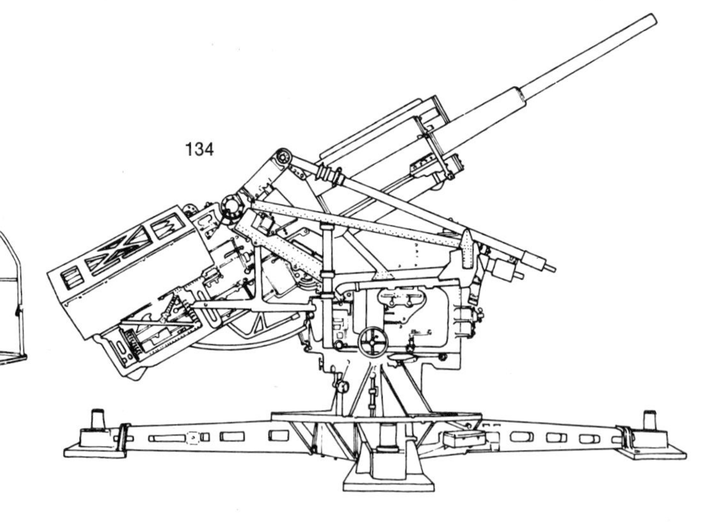

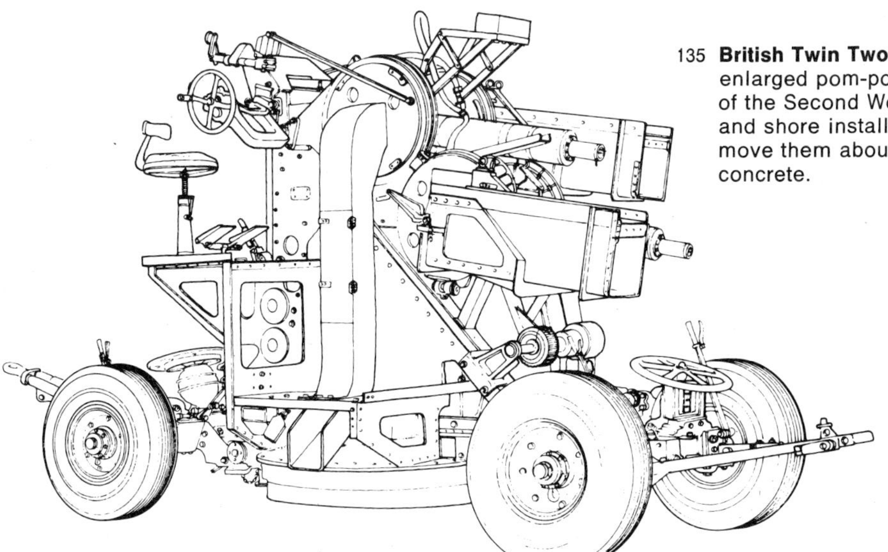

135 **British Twin Two-Pounder.** The two-pounder began as an enlarged pom-pom for use on board ships, but at the beginning of the Second World War a number were installed in dockyards and shore installations. A mobile mounting was provided to move them about, but in action they were emplaced firmly in concrete.

Instead of firing at a rate of twenty rounds a minute and employing predictors and tracking devices it would be necessary to increase the rate of fire and shoot by eye. To increase the rate of fire meant a smaller weapon to handle lighter ammunition, with a mechanism rather scaled up from a machine gun than scaled down from an artillery piece. The small dimensions of the ammunition ruled out time fuzes, since they could not be made to such a small size with any hope of accuracy in operation. The formula thus crystallised into a small automatic gun firing impact-fuzed shells, laid by direct vision, and capable of high speeds in traverse and elevation.

During the closing months of the First World War a German designer, Becker, had produced a 20mm cannon: an automatic weapon for use either as an infantry cannon or as an aircraft weapon. The patents for this gun were later bought by a Swiss company and after a certain amount of cleaning up and re-design it became the first of many Oerlikon 20mm cannon. Similar weapons were developed by Solothurn and Hispano-Suiza, and this type of cannon was soon touted as a possible light anti-aircraft gun. While they had the advantage of fast training, high rate of fire and so forth, their principal defect was the light weight of shell they fired, of which many were needed to ensure fatal damage to the target. Bofors of Sweden produced a 40mm automatic gun in 1929, and this was subsequently re-worked and improved to become the well-known 40mm Bofors of the Second World War (139). Germany, France and the USA all opted for a 37mm gun, though not of the same designs, while Germany hedged its bet by adopting the 20mm Solothurn cannon as well. All these weapons relied on simple visual sights, although as time wore on they began to sprout additional features, such as clockwork drives to set in target speeds and similar refinements. Periodically the sight would reach the point where the gunlayer had to have a degree in optics, at which there would be a reversion to something simple and the process would start again.

The Predictor was, by now, a fairly well-designed and understood piece of engineering, though it must be borne in mind that the predictor of those days was almost entirely mechanical in operation. The Sperry T8E3 which the US Army adopted in about 1937 consisted of no less than 3500 individual parts, excluding such things as nuts and bolts, and such machines were heavy and cumbersome to emplace. Data transmission was becoming commonplace, and as early as 1930 the US Army had a completely remote-controlled 3 inch AA gun, the pointing and elevating of which were done by servo-motors controlled from the predictor – all the gunners had to do was keep loading it. But such devices were not to become standard issue for several years.

By the middle 1930s radar had begun to be appreciated and steps were being taken to adapt it, first as an early warning system and, immediately thereafter, to link it with guns to provide a substitute for optical tracking. The development of radar has been thoroughly documented and recorded, and we cannot afford space here to study it. Suffice it to say that in 1939 Britain and Germany both had early warning systems in service; Britain had her first Gun Layer sets working with static artillery; and the US were beginning their early warning systems and also looking into gunlaying.

These years had also seen the polishing of heavy anti-aircraft gun designs until each of the future belligerents had made his choice and had started arming. Britain adopted a 3·7 inch gun firing a 28 lb shell to a 30,000 foot ceiling. This was an exceptionally good design, born of a combination of Vickers and Woolwich Arsenal ideas; adopted in 1936 it was still in service in some countries over thirty years later. The shell had a clockwork time fuze, though these were in short supply during the early days and a powder-burning pattern was still issued. For the protection of naval ports the Army adopted the 4·5 inch naval gun, for the sake of commonality in ammunition supply. This fired a 55 lb shell and was originally provided with a Swiss clockwork fuze, the Tavaro, though difficulties of wartime supply soon reduced stocks and rendered it obsolete, it being superseded by a British design similar to that used in the 3·7 inch.

Germany, of course, had the eighty-eight; this gun, adopted in 1935, was to become probably the most famous gun of the war; largely because of its anti-tank performance though, and not from any ascendancy in the anti-aircraft world. It was improved in 1937 by making it easier to mass-produce and adding an improved data-transmission system, and in this form saw the war out. With its 32 lb shell fitted with a reliable clockwork fuze it was a good, serviceable weapon, on a par with the British 3·7 inch.

The US had a 3 inch which was simply an improved model of the gun they had fielded at the close of World War One, which itself had been a high-angle adaptation of the 3 inch coast gun of 1903. There was room for improvement here, and a 90mm high-velocity gun of outstanding design was ready just before the US became involved in the war. A 105mm had also been under development off and on since 1919, and this came into service in 1940. And finally, a 120mm gun, also a brainchild of General Pershing in 1919, had been developed and was introduced just before Pearl Harbor, though this gun was not considered sufficiently manoeuvrable enough to be sent to field armies and was declared a home defence weapon. Be that as it may, four turned up some years back in a barn in Northern Ireland; perhaps some GI brought them over in his B Bag. Life is full of little mysteries.

The improvement in predictors and the advent of radar left the guns working on what was virtually the Central Post system, but now the size of the group varied. In some cases in Germany as many as 48 guns would be controlled from one command post, with one radar set for every six guns, all feeding their data to the central predictor and acting as back-up for each other in case one set went out of action. In Britain the principal target, London, was so close to the coast that it became necessary to construct special forts in the Thames Estuary and along the east coast. The first were manned by the Royal Navy and were pre-fabricated of concrete in dry-docks, floated out to their positions fully armed and manned, and there allowed to sink until they rested on the sea-bed, with the gun-floors well above the high-water mark. Each fort had two 3·7 inch guns, two or four 40mm Bofors and a radar. Then came the "Maunsell" forts, manned by the Army. These were constructed in separate units, floated out and sunk and then joined together with catwalks into batches of six platforms, four mounting 3·7 inch guns, one a Bofors for defence against low-flyers, and the central one carrying radar and command post. After the war some were demolished, some taken over as lightvessel replacements and signal stations, and some abandoned. Most people never knew that they had existed until the 1960s, when they were pitchforked into notoriety by their usurpation by the "pirate" radio stations.

Once the war was under way, a gap in the defences made itself apparent on both sides of the Channel. While light guns could cope with low-flyers, and heavy guns with high-flyers, what happened to the medium-flyers? The ceiling of the average light gun was about 6000 feet, so since the shells were point-impact fuzed, they had to be fitted with a self-destruction device to ensure that they went off before falling back to earth if they should miss the target, and so the gun's maximum range was governed by this feature. The heavy guns were limited in their effectiveness by the fact that they could not train and swing fast enough onto targets much below ten to twelve thousand feet – the feature of their construction which had led to the light guns in the first place. So there was, between six and ten thousand feet, a belt of sky in which the attacker was relatively immune. To fill up this gap it became vital to develop an Intermediate Anti-Aircraft gun; one which would have a greater range than the light gun, but faster pointing than the heavy, with a rate of fire somewhere between the two and with a reasonable-sized shell to compensate – since the calibre would still be too small for a time fuze – and a high rate of fire, or at least a large volume of metal in the sky, was the only valid method of attack.

Germany appreciated the gap first, well before the war, and in 1938 Rheinmettal were given a contract to develop a 5cm gun which became known as the 5cm Flak 41. A number were issued in 1941 to selected units in order that the idea of Intermediate AA could be explored and an opinion formed as to whether the idea was as good as it appeared on paper. For a variety of reasons – balance, roadworthiness and general difficulty of handling – the 5cm Flak 41 was not a particularly good gun, but it was good enough to show that Intermediate AA guns were a desirable property if the bugs could be ironed out. In view of this opinion work began in 1943 on a completely integrated weapon system in which there would be a six-gun battery with full remote power control, radar, predictor and displacement calculator all as a coherent fire unit. The gun got to the prototype stage before the war ended, but the whole system ran into innumerable snags and was never assembled. It is interesting to speculate on the fact that a 57mm Soviet AA gun introduced in the early 1960s shows every sign of descent from this German 5·5cm Gerat 58 experiment.

In Britain the gap was not appreciated until rather later, and work began in 1940; at this time every factory was hard at work producing standard weapons to replace the losses at Dunkirk, and new weapons were ruthlessly withheld until sufficient stocks of proven guns were available for defence. So there was no hope of producing a brand-new design and hoping to have it manufactured; the question was "What have we got that can be worked over to suit our purpose?" The first likely contender was the obsolescent Coast Artillery 3-pounder; this had the virtue that a new predictor recently designed for the 40mm Bofors could easily be adapted to it, but it was eventually turned down because the shell was insufficiently lethal. The next possible was the Coast Artillery 6-pounder of 6 cwt; the shell was suitable and work began on adapting it and providing some form of automatic loading. Early trials showed that a better performance was wanted, and a new barrel was developed, turning it into the 6-pounder of 10 cwt. A twin mounting on a three-wheeled trailer was produced and the Molins Company were asked to produce a suitable loading mechanism to step up the rate of fire. The original 6-pounder breech mechanism, a vertical sliding block, was used and attempting to turn this sort of mechanism into an automatic was a forlorn hope. The best that could be done was a mechanical feed and rammer system using large hoppers for the cartridges. In the event by the time the gun was far enough developed to have some hope of being turned into a service weapon, the Allied air superiority had removed the need for it. A few were retained as vehicles for testing auto-loading devices, but none entered service. So, one way and another, neither side ever managed to produce their Intermediate AA gun before the war ended. So far as is known, the Americans – no doubt made wise by the misfortunes of others – never entered this contest.

136 **German 5cm Flak SP.** Most people had given up mounting AA guns on trucks by the time of the Second World War, but the German Army had heavier trucks than most and found the idea particularly useful in protecting troop movements across the vast emptiness of Russia.

1941 saw a general improvement of the heavies for by now the targets were beginning to fly higher and faster. Germany produced the Rheinmettal design of 88mm Flak 41; apart from its calibre it bore little resemblance to the earlier Krupp model. The mounting was based on a turntable, instead of a pedestal, giving a much lower silhouette, and a larger cartridge case was used with a heavier charge to get greater ceiling and a shorter time of flight. It was this cartridge case which turned out to be the Achilles heel of the whole design, since it developed the unfortunate habit of sticking in the chamber and failing to eject cleanly; thus the empty case fouled the incoming round and jammed the breech. Many and varied were the expedients tried to conquer this fault, from recontouring the chamber to designing a special brass case (most German weapons were using steel cases by this time) but to little avail. The Flak 41 was like the little girl; when it was good it was very, very good, but when it was bad . . . . .

By this time 105mm and 128mm guns were in volume production. The 105mm provided the backbone of the static defences of Germany, being widely mounted in twin mounts on special Flak Towers in the major cities, and also in railroad trucks to provide special Flak trains which could be moved from place to place. The 128mm too, though in lesser numbers, was for static defence of important targets and was also provided in a twin-barrelled version for flak towers (133).

In Britain the search for higher performance was on by 1941, and several possible avenues were scouted; these included a squeeze-bore solution, a discarding sabot solution, and a 4·5 inch gun linered down to 3·7 inches.

137
The Airborne Bofors. This variation of the Bofors was developed in order to provide airborne units with a lightweight AA weapon. In this drawing the barrel has been removed for stowing on board an aeroplane.

138 **German 5.5cm Gerat 58.** This weapon was intended to be the finest medium AA gun in the world, forming part of an integrated weapon system, but it was begun too late and development was not complete when the war ended.

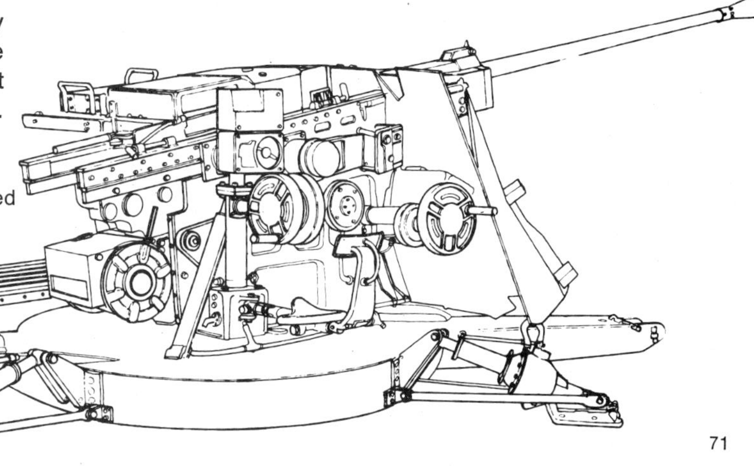

139
140
141

The idea finally adopted was virtually to fit a 4·5 inch gun with a 3·7 inch barrel; the chamber and cartridge were basically 4·5 inch design, thus giving a mighty punch to the 3·7 inch shell. In addition, to improve the ballistic performance, a special barrel developed by Colonel Probert of the Research Department Woolwich Arsenal, was fitted (144). In this design of barrel the rifling began in the normal way, but as the grooves passed up the barrel they gradually became less and less deep until some three feet from the muzzle they vanished altogether leaving the remainder of the gun smoothbored. To suit this barrel special shells were produced, having a thick and prominent driving band and two serrated steadying bands near the shoulder. In travelling up the special barrel the bands were smoothed down until the shell was ejected from the muzzle with no protuberances at all, a perfectly smooth surface which presented no obstacle to smooth airflow over it. The result of this was to allow the shell to get up to 35,000 feet in the time as the older 3·7 had taken to reach 25,000 feet.

To produce heavier metal in the sky to counter the larger and more robust aircraft, the British Army adopted another naval gun, this time the 5·25 inch. This was for static defence of ports and harbours in Britain, since it required quite a complex emplacement to provide full power operation for elevation traverse, hoisting ammunition and ramming. The round was a separate loading, shell and cartridge unit, with the shell weighing 80 lbs. The original naval version taken over was a single-gun model but the Army developed a twin mounting, though this did not enter service in great numbers until after the war was over.

In Germany too the trend was towards a bigger bang at a higher altitude, and contracts went out for a 15cm and even a 21cm superheavy AA gun. These were to be full-powered static mountings with magazine-loading to give a rapid burst of fire without the disadvantage of having to feed heavy rounds during an engagement. Prototypes were built, but neither weapon ever came into production, since by this time – early 1944 – German designers were already looking forward to the guided missile and tending to discount the gun as being a primitive device, the first of many people to make the same mistake.

The launching of the pilotless V-1 bombs in 1944 was the anti-aircraft gunner's finest hour. Here was a target which was performing just the way the predictor liked them –

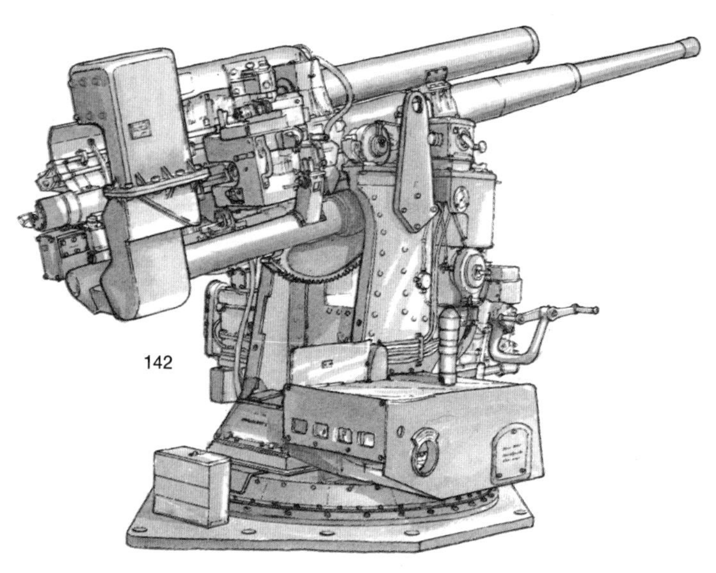

139 **The Bofors Gun.** The universal 40mm Bofors, used by almost every combatant during the Second World War, is one of the all-time classic designs.

140 **German 88mm Flak 41.** The Rheinmettal design of improved '88', developed in order to obtain a higher ceiling and velocity. Turntable mounted it proved to be an even better weapon than its predecessor, once its initial development difficulties had been overcome.

141 **British 3.7 inch Mobile.** Introduced in 1937 this became the standard British AA gun and remained in service until the 1950s. Later models incorporated power ramming and automatic fuze setting.

142 **British 3.7 inch Static.** As well as being a mobile gun for use with field forces the 3.7 inch was widely emplaced in fixed defences throughout Britain. Since it could be heavier than the mobile version, a large counterweight replaced the equilibrators and balanced the barrel weight.

143 **German 88mm Flak 36.** The famous German 'Eighty-Eight' in its original anti-aircraft guise. It later achieved more fame as an anti-tank gun, but it remained the backbone of Germany's air defence throughout the war.

143

flying straight, level and at a steady speed. Though this delightful picture was somewhat marred by the fact that the combination of height and speed chosen by the designer of the V-1 made it a target which scooted across the sky at an alarming rate of change for the poor gunlayers. The long-sought Intermediate guns would have proved their worth here, but in the event it was the heavies which gave the best results. In the initial stages of the battle the guns were deployed thickly on the outskirts of London, since this was the major target – the Royal Air Force working outward from there. This soon showed itself to be the wrong approach, since shooting the "doodlebugs" down often failed to detonate them until they hit the ground, and the subsequent damage was often as bad as if they'd been left alone. Very rapidly the guns were uprooted and re-deployed in a belt along the coast. British units were interspersed with US Army 90mm gun units, together with 40mm Bofors from British and US units, 20mm cannon of various sorts and even a number of .50 Browning machine guns on multiple mountings. After a few days the .50 and 20mm weapons were withdrawn, since they were little use. It has been said in an official report that after two days of redeployment "any unit arriving late would have found it hard to fit its guns in" since almost every available piece of open ground was armed with an anti-aircraft weapon of some sort.

This grouping showed an immediate effect, and the number of V-1s which survived was a very small percentage of those launched. After a few weeks the US Army 90mm guns were withdrawn for service in Europe, and the whole defence now devolved on British 3·7 inch guns. New developments in fire control equipment, notably the American SCR 584 Radar and its associated predictor, were installed to deal with the speed and height requirements of these peculiar targets, and travelling teams of instructors kept the gunners fully conversant with each new device. One of the most revolutionary and effective devices which came into use about this time was the Proximity Fuze (159).

Time fuzes rely on the user estimating the time of flight to the target and then setting this on the fuze prior to loading; this demands an accurate assessment of the target's future position, accurate setting and an accurate mechanism which will detonate the shell at exactly the set time, with the minimum of tolerance. If all these things are correct, then the shell will burst within lethal distance of the target and do the desired damage. But the chances of everything going well, of having an accurate assessment, accurate setting and accurate mechanism are remote. The proximity fuze relies on the fuze being able to detect when it is within lethal distance and then, without further ado, detonating the shell – in other words it reacts to the proxi-

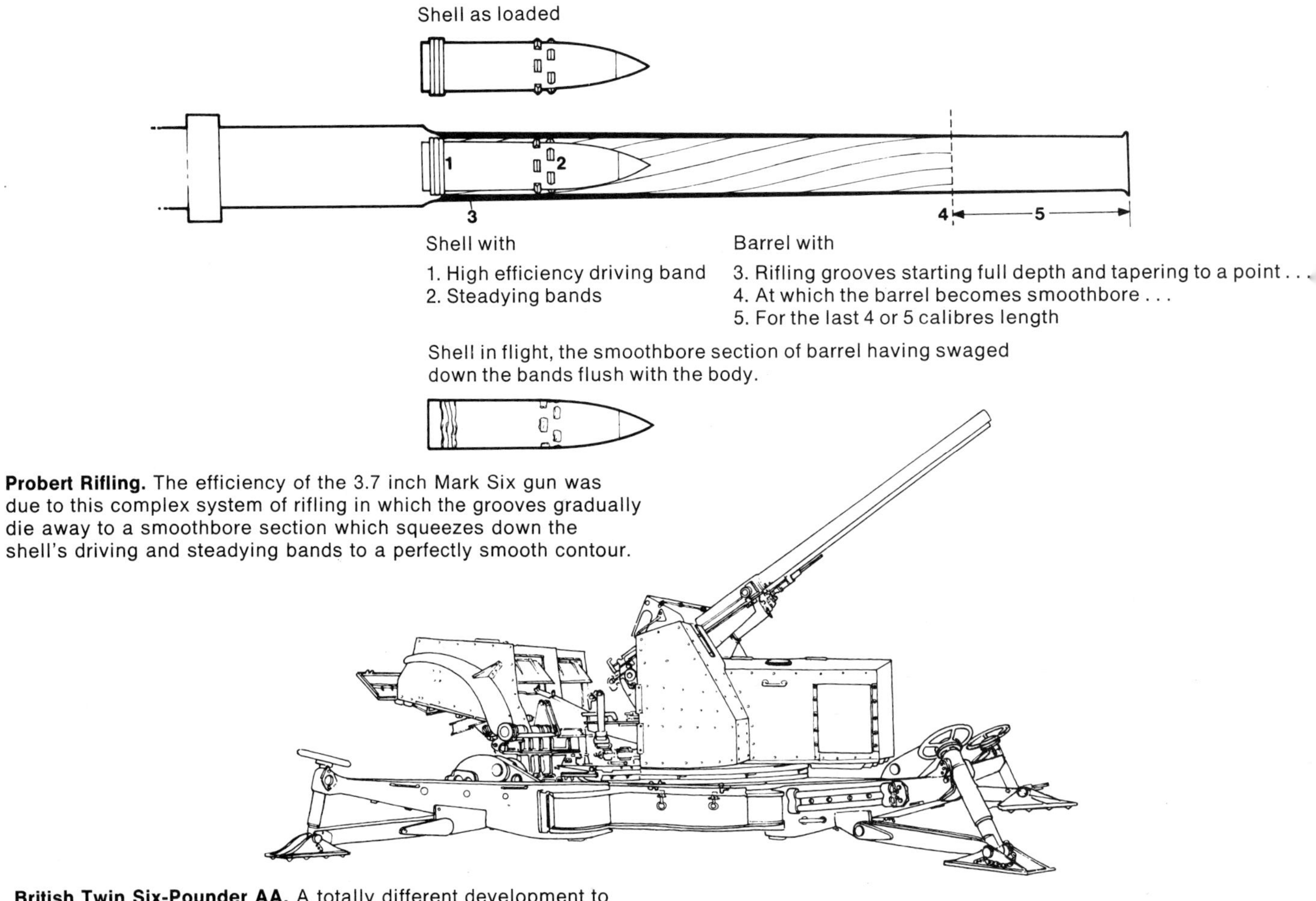

144 **Probert Rifling.** The efficiency of the 3.7 inch Mark Six gun was due to this complex system of rifling in which the grooves gradually die away to a smoothbore section which squeezes down the shell's driving and steadying bands to a perfectly smooth contour.

145 **British Twin Six-Pounder AA.** A totally different development to the successful twin six-pounder coast gun, this used the same barrels, mounted them on a three-wheeled trailer, and grafted a complex loading mechanism on behind. The results were not in proportion to the effort put in.

mity of the target. The idea is not new; inventors were playing about with it in the 1930s and there was an effective proximity fuze in British service in 1941, but it was for use in rockets. Rockets have a lower acceleration than shells, and the electrical circuitry was only capable of withstanding rocket thrusts, not the violent blow given to a shell when it is fired from a gun. This early model worked on photo-electric principles; when the shadow of the aircraft fell across windows in the fuze, it detonated. Unfortunately it was only of use during daylight, and when the bulk of air attacks came at night, it fell into disuse.

When radar was well under way, some of the workers in this field postulated a fuze which would detect the slight reflections of electrical energy from the aircraft target due to it being "illuminated" with a radar beam, but the amount of energy reflected is so small that no circuit capable of being compressed into a workable fuze would ever detect it. Then the idea of producing a self-contained transmitter and receiver within the fuze took hold. The theory seemed sound, the practical aspects were worked out; but then the problem of manufacture arose. In Britain at that time the radio and electronics firms were working flat out on radar and radio equipment, and no room could be spared for the proximity fuze. So it crossed the Atlantic with the Tizard Committee and was given to the US Navy. They in turn contracted it out to the Eastman Kodak Company, Sylvania (for the tiny valves) and Exide (for the tiny but powerful batteries needed). All the manifold problems appeared and were solved, and the USS *Helena*, in June 1943, had the honour of firing the first proximity fuze in action, in the South Pacific. With slight modifications to suit British shells, they were then produced in quantity in time for the flying-bomb campaign, and were later provided for field artillery to provide airbursts above the ground for anti-personnel effect, this application being first used in the Ardennes in December 1944.

The proximity fuze as finally perfected consists of a plastic head containing a radio transmitter, receiver and aerial. Below this, in the elongated body, lies the battery – a selection of electrical and mechanical safety devices to insure against premature functioning – and the fuze magazine. When the shell is fired the sudden shock activates the battery, which provides power to operate the two radio circuits. A signal is sent out through the aerial, and the area covered by the signal is carefully matched to the lethal area of the shell, so that anything which reflects the wave will be in a position to receive the subsequent benefits. A reflected signal comes back from the target into the aerial, and the interaction between incoming and outgoing signals sets up a difference frequency which is detected and when its strength indicates that the target lies well within lethal distance, this frequency triggers a firing circuit which detonates the fuze magazine and thus the shell.

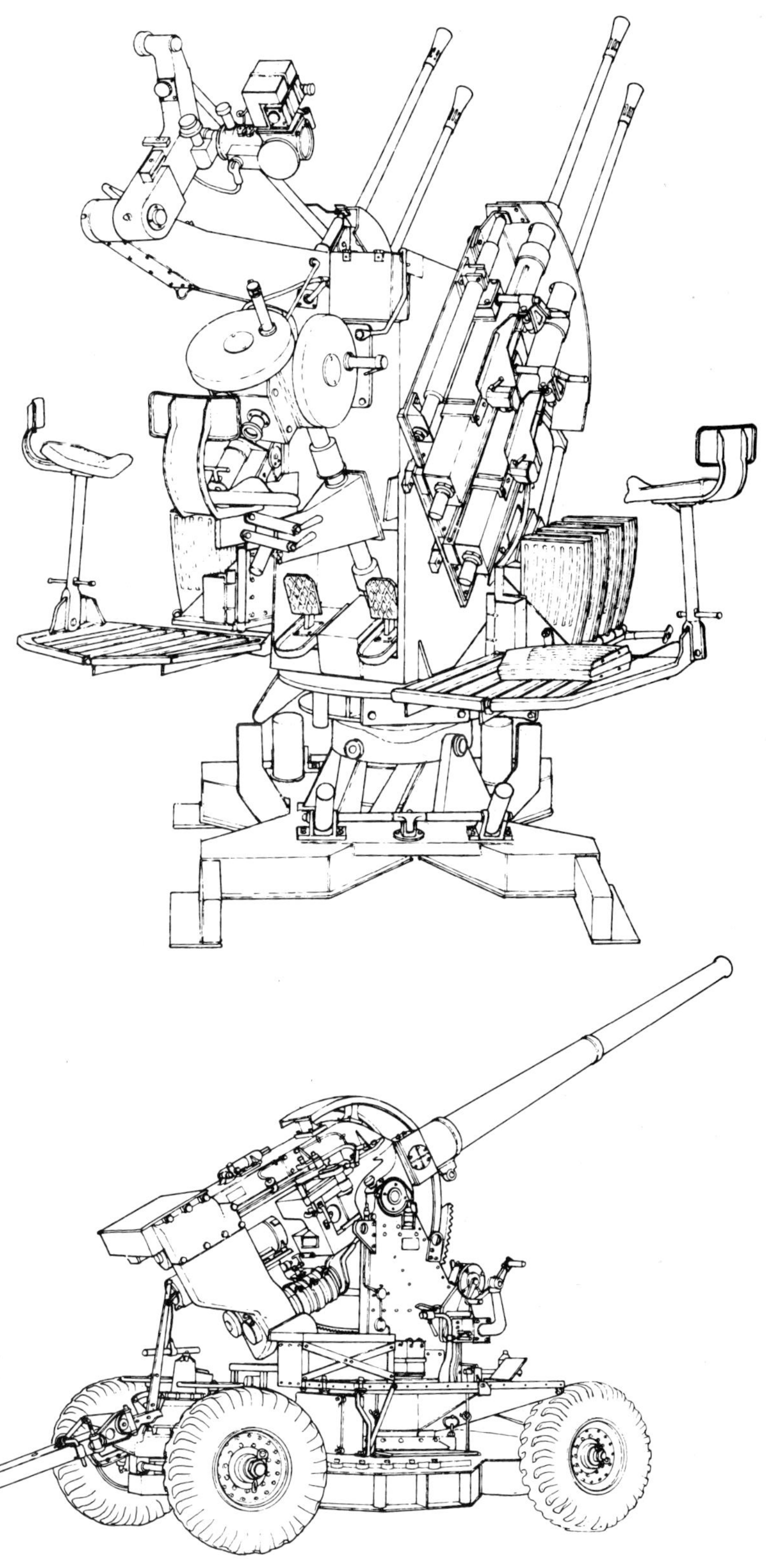

146 **German 2cm Flakvierling.** Another method of filling the air with metal is to use a small calibre fast-firing multiple weapon. This four-barrelled cannon was a standard German Army weapon for defence against low-flying attacks.

147 **British 3.7 inch Mark Six.** To improve the performance of the 3.7 inch gun the 4.5 inch was lined down to shoot the 3.7 inch shell using a 4.5 inch cartridge. It proved to be a most successful weapon and saw the heavy gun era completely out, being replaced by a guided missile.

One of the disadvantages of the early types was that should the shell pass close to a friendly force in its flight – such as crossing over a hill top held by one's own forward troops – it would detonate. But this has been cured in recent years by the addition of a simple timing clock which delays full operation of the circuits until the shell is well away from friendly troops. In the case of anti-aircraft fuzes a self-destruction unit was also fitted that ensured no shell could return to the ground in a live and whole condition. For anti-aircraft gunners, the principal advantages of the proximity fuze were that the calculation of the fuze length no longer had to be done and the setting of the fuze was also no longer part of the loading drill, all of which speeded up the rate of fire as well as making the chances of a burst in lethal distance more likely. The proof of the pudding was in the eating: when the final figures of the flying-bomb campaign were calculated, it was found that 94 per cent of the downed bombs were victims of gunfire.

When the Second World War ended it was "back to the drawing board" in Britain and the USA. Increased speeds and heights of aircraft with the introduction of jet and rocket propulsion meant more powerful guns were needed, and, more important, guns with a higher rate of fire since engagement times were going to be shorter than ever. Since the proximity fuze omitted the fuze-setting step from the loading cycle, the US had begun, before the war ended, on the design of a gun, which taking advantage of this omission in drill, was to be an all-proximity fuze weapon. Its calibre was 75mm, this being selected as the smallest calibre which was still sufficiently lethal when the over-sized proximity fuze was fitted, displacing some of the high explosive. Two equipments were planned, the T18 to be a short-term project with fire control as a separate unit and the T19, a long-term project with the fire control assembly built into the carriage. The T18 was soon abandoned when the war was seen to be drawing to a quicker close than expected, and the T19 was worked on until it was eventually issued to service in the early 1950s as the M51 "Skysweeper" (155). It carried 20 rounds in twin revolver-pattern magazines in rear of the gun, and had all radar and fire control equipment integral with the gun mounting. It was the USA's last anti-aircraft development, and probably the best of its kind ever seen.

Britain began, in its usual atmosphere of post-war penury, to improve on existing guns; the 3·7 inch Mark 6 (147) – the Probertised gun – had all the performance needed to

148 **US 90mm AA Gun.** The standard US AA gun of the Second World War was the 90mm. An improved mounting was later developed which enabled the gun to be used also as an anti-tank and anti-torpedo-boat gun.

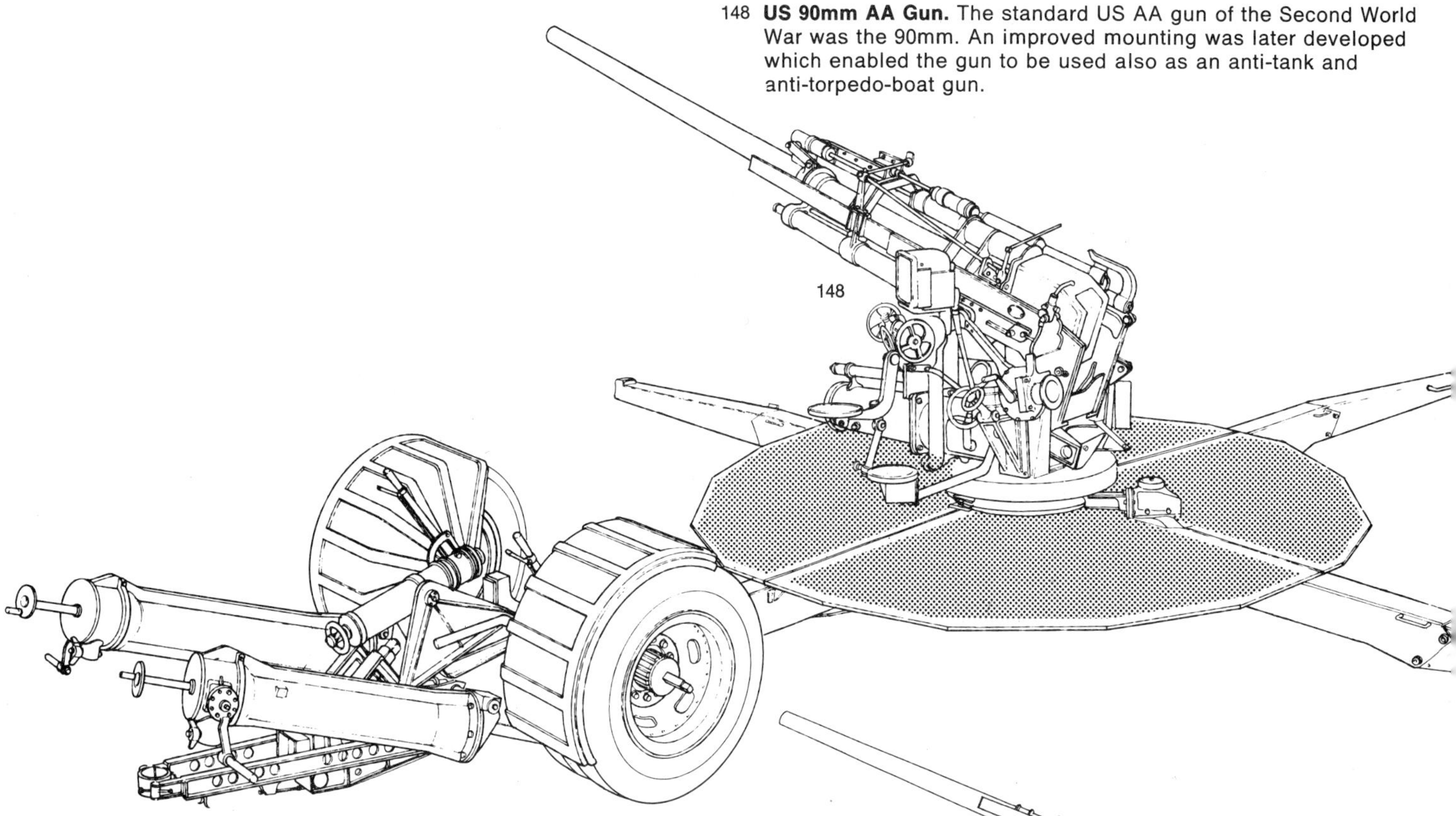

148

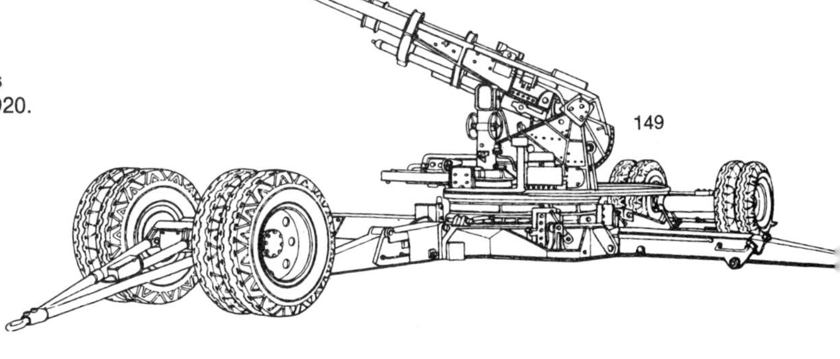

149

149 **US 120mm AA Gun.** The heaviest US AA equipment with this 120mm, the descendant of the Pershing-inspired model of 1920. It could cope with the highest fliers of its day but was rarely seen outside the USA.

150 German 2cm general purpose gun mounted on a half-track for manoeuverability.

151 Swedish 75mm Anti-aircraft guns in action. Photo Bofors A.G.

150

151

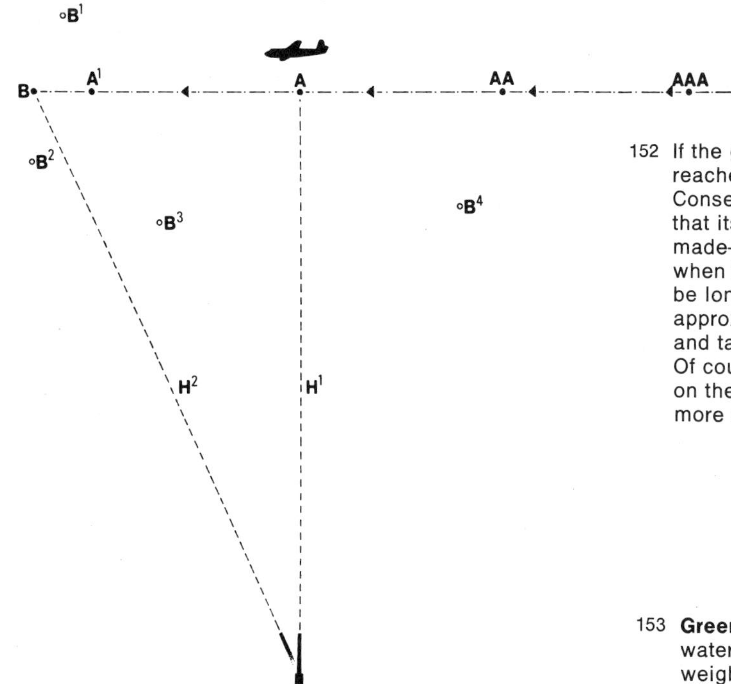

152 If the gun fires when the target is at A, by the time the shell reaches A the target will be at A^1, having travelled D in the time. Consequently the plane must be tracked from AAA to AA and A so that its course and speed can be measured and a prediction made—if the gun is fired when the plane is at A, it will be at A^1 when the shell reaches A. If we point at A^1, the time of flight will be longer, since gun-A^1 is greater than gun-A. By a series of approximations the point B is determined as the point where shell and target will coincide and the gun data and fuze set accordingly. Of course, if the pilot decides not to continue at the same speed on the same course, he could be at B^1, B^2, B^3, B^4 or many more options when the shell arrives at B . . .

153 **Green Mace.** Britain's last AA development was this 5 inch water-cooled magazine-fed gun, firing at 96 rounds a minute and weighing 30 tons on the move. It was rendered obsolete overnight by the guided missile.

cope with aircraft for the following decade, provided that the rate of fire could be significantly improved. In 1946 the "Ratefixer" programme began, which was the application of fresh minds to the problem of loading a gun as fast as possible. Without going into abstruse theories, it can be said that previous thought on loading was based simply on speeding up the basic manual movements; now careful analysis showed that if the paths of the entering round and the exiting cartridge case were carefully planned, it might be possible to start loading before unloading was completed, thus telescoping the various movements into a shorter cycle. Several designs were developed, using belt, magazine and other types of feed, and given thorough trials.

The programme worked well and eventually produced rates of fire of up to 96 rounds per minute – this with a 3·7 inch gun and a sixty-five pound cartridge, but advances in aircraft design had been quicker than anticipated, and by 1950 it was seen that the gun would soon be out of date and a completely new equipment was needed. Soon a 4·26 inch/3·2 inch taper-bore gun was proposed and also a 5 inch gun firing a fin-stabilised dart projectile, both weapons to go on the same mounting as alternatives.The mounting was to have twin revolving magazines feeding alternately a total of 28 rounds; the barrels were to be water-cooled; the magazines were to be capable of quick-changing for rapid re-loading, and the whole mounting had to be mobile and fully remote power controlled.

Within a short time it was found that the taper-bore gun was going to be a stiffer proposition than at first thought, and, as a hedge against delays, a conventional 102mm gun was produced to fit the mounting. The whole project became a little confusing, and the name "Green Mace" though strictly applicable only to the 5 inch barrel project, came to be loosely applied to whatever gun happened to get on the mounting. Further delays bedevilled the project,

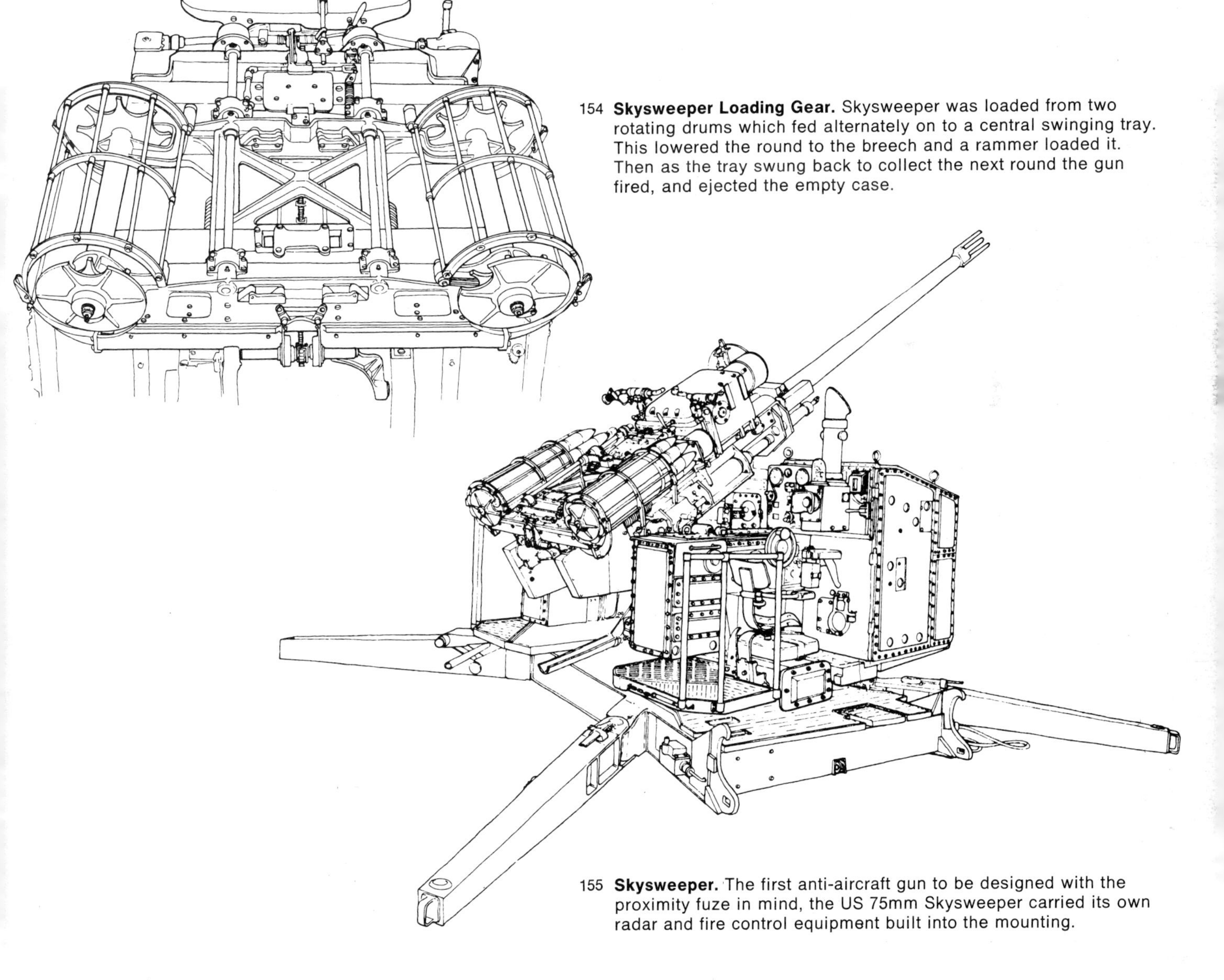

154 **Skysweeper Loading Gear.** Skysweeper was loaded from two rotating drums which fed alternately on to a central swinging tray. This lowered the round to the breech and a rammer loaded it. Then as the tray swung back to collect the next round the gun fired, and ejected the empty case.

155 **Skysweeper.** The first anti-aircraft gun to be designed with the proximity fuze in mind, the US 75mm Skysweeper carried its own radar and fire control equipment built into the mounting.

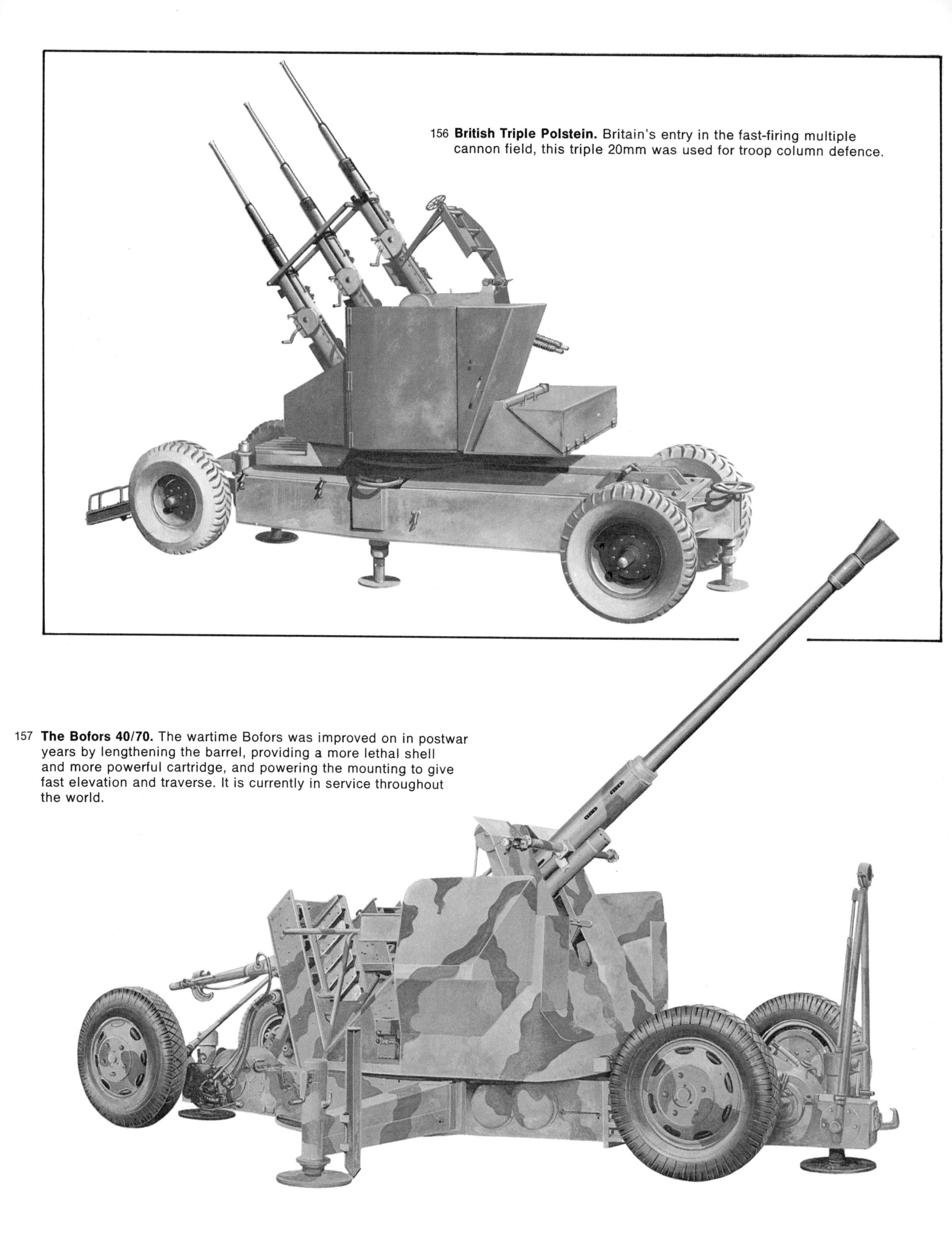

156 **British Triple Polstein.** Britain's entry in the fast-firing multiple cannon field, this triple 20mm was used for troop column defence.

157 **The Bofors 40/70.** The wartime Bofors was improved on in postwar years by lengthening the barrel, providing a more lethal shell and more powerful cartridge, and powering the mounting to give fast elevation and traverse. It is currently in service throughout the world.

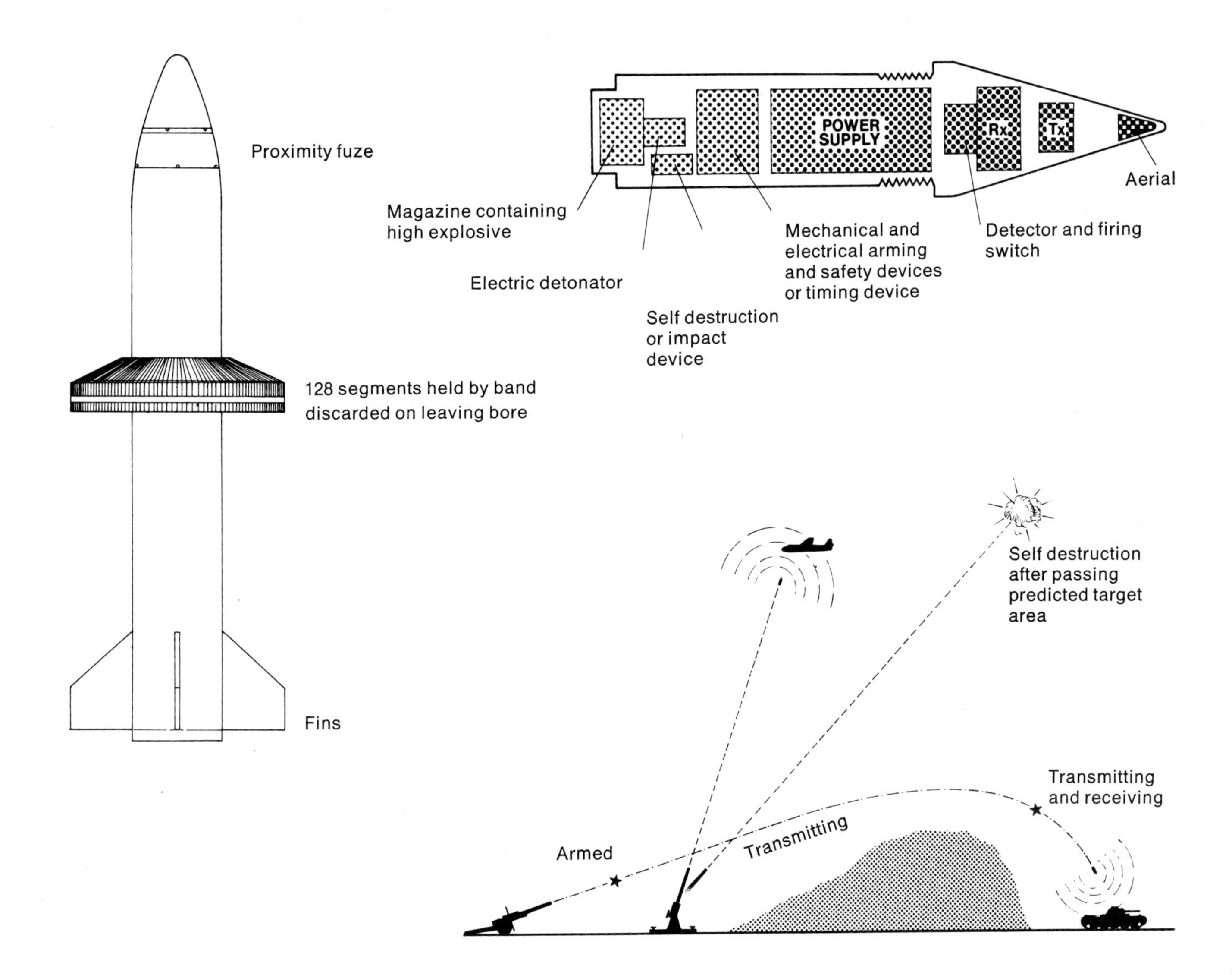

158 **Discarding Sabot Shell.** The 5 inch dart shell for the British AA gun development 'Green Mace'. The central sabot steadied the shell in the bore and broke up into 128 discards on leaving the muzzle.

159 **The Proximity Fuze.** The proximity fuze uses a self-powered radio to detect the target and set itself off at the crucial moment; if the fuze misses an aircraft it will destroy itself. It is also used in the ground role when, to prevent false indications from terrain features, it does not become active until shortly before reaching the target

and by the mid-1950s, despairing of ever obtaining an anti-aircraft gun, the Army produced "Longhand", a reversion to the 3·7 inch Mark 6 with a 12-round rapid loading conveyor, by Ratefixer, out of Green Mace, Heath Robinson up. But although approved for service, none were ever made; the sands were running out rapidly for anti-aircraft gunnery.

In 1954 the 102mm version of Green Mace was fired very successfully, giving a rate of fire of 96 rounds a minute. In 1956 the 5 inch barrel was ready and firing trials again showed successful results. In 1957 the whole project was dropped, the guided missile having finally proved itself a workable proposition for anti-aircraft defence. The 102mm Green Mace is now in the possession of the Imperial War Museum; a 5 inch projectile is in the Royal Artillery Museum at Woolwich. Other than these, nothing remains of the strong arm which broke the V-Weapon attack in 1944.

In the light anti-aircraft world though, the gunner is still using his traditional tool, though for how long this state of affairs will last is not clear. After the war limited trials were carried out with the aim of improving on the 40mm Bofors; a 57mm version and a 42mm weapon called "Red Queen" were tried in Britain, but, as with many other countries, experiment was abandoned when A. B. Bofors produced their L/70 version of the wartime 40mm gun. With the rate of fire doubled, and with exceptionally efficient power traverse and elevation, a more powerful cartridge and more lethal shell, and backed up by a variety of sophisticated fire control systems, the L/70 could more than hold its own against low-flyers. But in the last few years the missile makers have begun to boil their ideas down into equipments small enough and reliable enough possibly to supplant the gun. We shall live and learn.

Anti-tank guns

160

First World War 37mm Anti-Tank Gun. The first artillery piece specifically designed to combat tanks was this little-known German 37mm gun. Very few were made and its velocity was too low to deal efficiently with the post-war tanks, which began to use better armour than the wartime models.

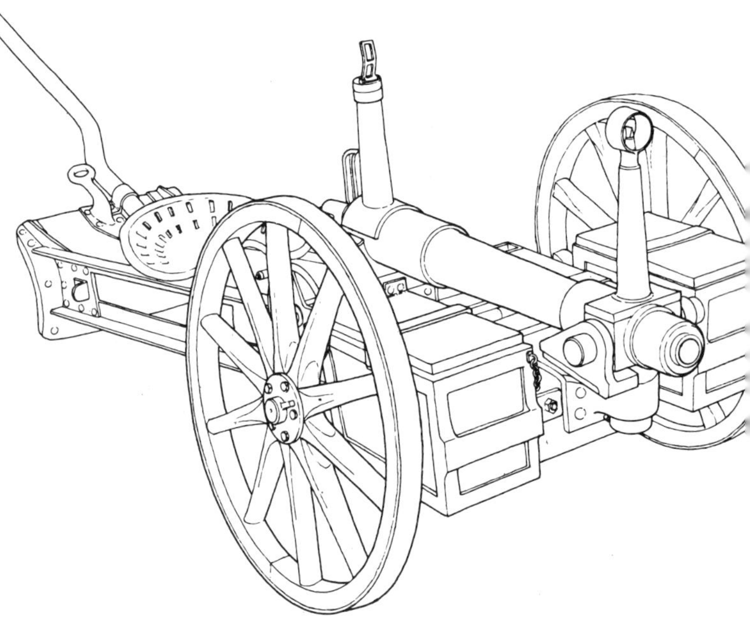

In 1916 the tank made its first hesitant appearance on the battlefield, and while the records of the time are replete with tales of how this new weapon struck fear and trembling into its opponents, the same records also bear out the fact that the German gunners soon got over their initial shock and did the best they could with the guns and ammunition at their disposal. Since those first tanks were constructed of relatively thin plate, a standard high explosive shell could do sufficient damage to put the tank out of action, given a fair hit, but the field guns were usually too far back to act as effective anti-tank weapons, since the tanks were into the infantry positions and causing havoc before getting close enough to the guns to be engaged with direct fire. The answer to that, of course, was to take guns forward and site them to cover likely tank approaches either giving the guns a wide field of fire or siting them to cover obstacles so that the oncoming tanks would be channelled to the gun, a technique which formed the basis of anti-tank tactics thereafter.

But a field gun deployed as an anti-tank gun is a field gun robbed from its primary role, a fact rediscovered in the Second World War, and towards the latter end of the war the Germans, as the principal sufferers, attempted to develop a specialised anti-tank weapon. Their efforts diverged into two distinct branches; firstly a one-man infantry weapon, and secondly a crew-served artillery-type weapon. The infantry were issued with the Mauser anti-tank rifle (161), an over-sized super-powered version of the standard Mauser bolt action rifle firing a 13mm bullet at high velocity. With a hard steel core this bullet could cut through the contemporary tanks like the proverbial knife through butter, and it fathered a generation of shoulder-fired anti-tank rifles built by various nations which remained in service until the latter part of the Second World War.

The artillery-type weapon was a simple 37mm cannon of low silhouette and wide traversing ability in order to cover a good field of fire from a concealed position (160). By latter-day standards this gun was hopelessly under-powered, its short barrel allowing only a modest velocity to be obtained, but even so it was adequate at the time, though few were made before the armistice and fewer still saw action.

After the war the two great discussion points among artillerymen were firstly how to combat aircraft and secondly how to beat the tank. Very soon the latter question died away as it became tacitly understood that tanks were expected to fight tanks, and what was left would be the responsibility of the infantry. Consequently the anti-tank problem was left in the hands of the tank people, with any crumbs of wisdom they dropped being snapped up by the infanteers. On the Continent they looked with favour on Becker's 20mm infantry cannon, which had made a slow start as a trench gun in 1917. Mounted on two wooden wheels it was a miniature field gun, but fired semi-automatically from a magazine. The shells were pointed, base-fuzed and filled HE or nose-fuzed HE for anti-personnel work. The piercing projectile was quite effective against the current tanks, i.e. the tanks left over

161 **Mauser Rifle Model 1918.** Since the early tanks were relatively thin this 13mm high-velocity shoulder rifle could penetrate their plates and deal with the occupants. This was the first of a long line of imitators, some of which served in the early years of the Second World War.

162 **British 18-Pounder on the Hogg & Paul Platform.** One problem confronting field guns in the anti-tank role was that of traversing to keep up with a moving target. Major Paul and Captain Hogg of the Royal Artillery developed this idea in 1918 and it was the progenitor of the platform used with the later 25-pounder gun.

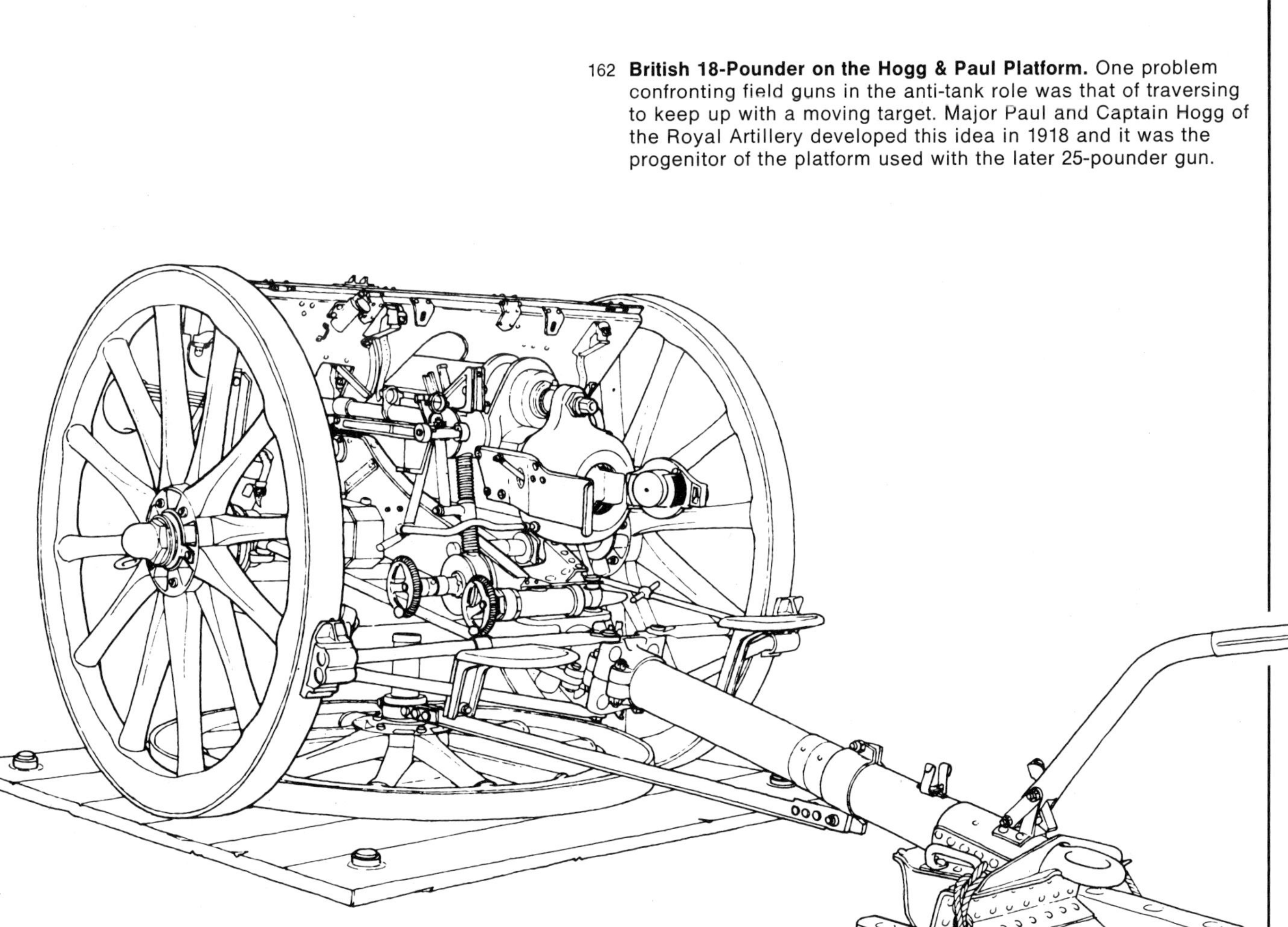

from the war, and several nations tentatively adopted the Becker in small numbers for extended trials.

Britain purchased a number of 20mm Solothurn cannon from Switzerland and, on a variety of wheeled and tracked carriages, these were extensively tested. In an attempt to develop a native weapon of comparable performance, the ·8 inch Elswick gun was developed, but neither this nor the Solothurn were taken into service since it was considered that their efficiency was little better than the ·55 inch Boys anti-tank rifle under contemporary development for the infantry and the task would be better performed by a heavier weapon. Throughout the world the same sort of opinion seems to have been held and the late 1930s saw a general trend to small, light, easily-handled weapons on two-wheeled carriages which could be manhandled into place by a couple of men, dragged about by them if necessary, and easily concealed. Britain had decided on a 2-pounder of 40mm calibre, Germany adopted the excellent Rheinmetall design of 37mm. Russia bought the German gun and America and Japan copied it. France pinned its faith on the 25mm Hotchkiss, and again Japan, having an each-way bet, followed suit. The Hotchkiss fired a solid bullet for piercing and an explosive bullet for anti-personnel effect. Germany, Russia, Japan and the USA went for the armour-piercing shell, and Britain, after a brief skirmish with the shell opted for the use of Armour Piercing Shot instead.

This brings us to the greatest controversy of all, one which raged for many years – that of shot versus shell. The AP Shot is a solid bullet of steel, carefully heat-treated and hardened to give a hard but brittle point, resistant to compression stresses, and with a graduated heat-treatment of the body so that the portion in rear of the point is less brittle and more tough, enabling it to resist side shearing stresses. These stresses inevitably occur when shooting at armour, since the odds on getting a shot at exact right-angles to the target, and thus confining the stress to a straight compression down the line of the shot, are pretty astronomical. Assuming the shot penetrates at whatever angle it strikes, it then punches out a slug of metal from the armour, which is driven into the tank at high speed, and passes through into the tank to bounce around inflicting grave damage on anything it meets. It is hoped that the shot will then follow through the hole and ricochet around inside to compound the mischief.

The AP Shell, on the other hand is not merely a solid bullet of steel, but is hollowed out to carry a charge of high explosive and a base fuze. Externally it is much the same as the shot, and its production and heat treatment follow similar lines, for after all it is going to be asked to penetrate the same sort of target. The intention is that the shell should pierce the armour by the kinetic energy of its arrival, punch out a similar plug to the shot, pass into the tank whole, and then have its explosive payload detonated by the base fuse. Thus the detonation and subsequent fragmenting of the shell will give a considerably enhanced lethal and damaging effect inside the tank.

On the face of it the shell seems to have indisputable advantages, but there are one or two drawbacks which become apparent on closer study. Firstly, by boring out the body to provide space for the explosive, the mass is reduced and therefore the penetrative effect is slightly impaired. The cavity also tends to weaken the shell's resistance to side-stress during penetration. Secondly, the selection of a suitable explosive is difficult; since it is intended to slam the projectile as hard as possible against an armoured target, selecting an explosive of high sensitivity means that it is liable to detonate from sheer astonishment as soon as the impact occurs. Selecting a relatively inert explosive, while overcoming the impact problem, means that detonating it at the right time may be difficult and also that its shattering effect on the shell might be poor. This impasse is generally overcome by using a powerful explosive, desensitising it by the admixture of wax, and providing an inert pad at the front end of the cavity to act as a shock absorber.

The fuze is the next problem area. The shell's operation demands that the fuze be activated by the impact shock but its initiation of the explosive must be delayed until the shell has passed through the plate and entered the tank. The simple solution is to put a pyrotechnic delay in the fuze, but obviously this is only applicable to one given thickness of armour and one striking velocity. Under these optimum conditions the shell will detonate in the right place but if the armour is thicker or the velocity lower, the detonation might well occur as the shell is still passing through the plate. Many designs of "thinking" fuzes have been put forward, in which the impact shock unlocks various safety devices and arms the fuze and the actual initiation is done by a mechanism which senses the change in deceleration as the shell pulls clear of the armour after penetration, but very few have ever been successfully applied, and most nations have stuck to simple delay fuzes, in the hope that they will provide the right answer most of the time.

The other problem with the fuze is its secure attachment. During penetration the hollow shell may be pinched or flexed, and the fuze can be popped out like a cork from a bottle, so that when the shell finally arrives in the tank it has no fuze there to initiate it.

In spite of all these objections most people went for the shell in the hopes that the drawbacks could be gradually designed out, because of the allure of the detonation. In fairness it has to be admitted that most nations did eventually produce a fairly satisfactory piercing shell for anti-tank use, and indeed the Soviets pinned their faith on such projectiles until the 1960s. The Germans were of the opinion that shot was wasteful since it usually went straight through the tank and out on the far side, doing little damage unless it struck a crew member or vital component during its traverse of the tank interior. Britain had developed AP shells for the 2-pounder when the gun was a tank weapon, and, after all, like most nations she had a legacy of piercing-shell design from coast and naval artillery, but the difficulties in such a small calibre were not thought to be worth struggling against, not for the sake of carting two or three ounces of explosives to the target and then probably losing its effect when it got there. AP Shot was the selected British projectile.

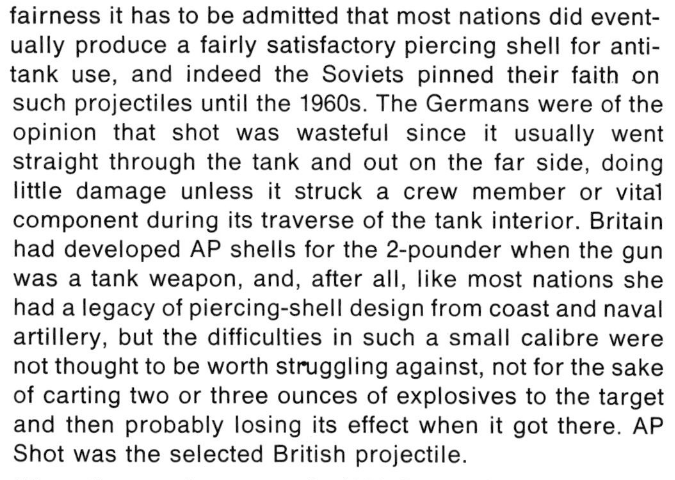

When the curtain went up in 1939 the combatants, and the

bystanders, were surprisingly alike in their anti-tank armament, deploying the guns outlined above, all around 37mm calibre, all firing a projectile of 1.5 to 2 pounds, all on light two-wheeled carriages. Britain, once again, stood aloof from the crowd, electing to man its anti-tank guns with artillerymen, while everyone else gave them to the overworked infantry.

The infantry of most nations were also provided with anti-tank rifles, descendants of the original Mauser design. These, although looked at askance today, were effective in their time, usually firing a heavy-cored bullet of about half an inch calibre at velocities in the 3500 f/second area. They were capable of penetrating side armour of the contemporary tanks, but they were unpleasant to fire and were soon overmatched by advances in tank design.

Poland, the Phoney War, the 10th of May 1940 came and went; the lesson learned throughout was that heavier tanks were coming and heavier weapons were going to be needed to cope with them. The British Army left 384 2-pounder guns on the beach at Dunkirk, an appalling percentage of its total holdings. In 1938 a replacement, a 6-pounder gun, had been proposed, and design was progressing in a desultory fashion. Now it was shelved; production of this gun would have jeopardised production of the 2-pounder, and a 2-pounder in the hand was worth a dozen untried 6-pounders in the production-modification-remodification pipeline.

The German Army had also begun work on an improved anti-tank gun in 1938, this one a 5cm weapon, and at the same time began to hedge its bet technologically by developing a totally new kind of weapon – the squeezebore gun.

The squeezebore was not a new idea; one Karl Puff had patented it in the early 1900s but nobody had bothered to follow it up, since Puff, in effect, had said "Here is the idea; how you go about making it work is your problem". The principle he expounded was that if you make your gun barrel taper down to a small calibre as it approaches the muzzle, then since the base area of the projectile is decreasing more or less proportionally with the expanding propelling gas, you obtain an increase in velocity.

Puff's idea was eventually taken up by another German engineer, Gerlich. His first attempts were with small arms and he spent the 1920s perfecting a taper-bore hunting rifle. He then tried to interest various military forces in the taper bore rifle as a sniper weapon, but failed to sell the idea. Then he turned to the anti-tank problem. By making a shot with two collapsible flanges on the body, it was possible to swage the flanges down during shot travel, giving

163. **The Becker-Semag Cannon.** The fore-runner of the Oerlikon gun, this was intended as a light 20mm infantry support cannon, but many nations tested it out as a potential anti-tank gun in the 1920s.

desired high velocity together with a flat trajectory. In order to let the shot deform properly, it was necessary to make it with a core and apply the flanges as separate pieces. If the core were of steel, the velocity attained was so high that the core would shatter without penetration. This phenomenon was nothing new; it had been found years before when dealing with plain steel shot, and a method of avoiding it was to place shaped caps on the shell (112) which would spread the stresses more evenly over the head of the shell instead of concentrating the shock of impact into the point. But even the addition of caps was only capable of deferring the shatter to a slightly higher striking velocity; the striking velocity of Gerlich's taper shot was far beyond anything which had been tried before, and there was no hope of a steel projectile arriving at such speeds and surviving. The only solution was to use a substance harder and more resistant to shatter than steel, and the only practical substance is tungsten carbide. Diamond hard, this will resist shatter to any velocity capable of being reached by conventional guns, and this was what Gerlich selected. By making a core of tungsten and surrounding it with a thin sheath of steel to carry the flanges, he finally arrived at a highly effective anti-tank projectile for his taper-bore gun.

Thus the German Army were issued with the 2·8cm Schweres Panzerbuchse Model 41 (167), which started out as a 28mm gun at the breech, but expelled a 21mm projectile at over 4000 feet per second. This was first unveiled in Libya, where the first British 6-pounder guns were at last being issued, production having finally begun. The Americans, keeping a watchful eye on the world's battlefields, realised that their 37mm was by now outdated, and started looking for a heavier weapon. Factories in the USA and Canada were by now making British 6-pounder ammunition on contract, so the US Army astutely adopted the British 6-pounder; originally built to British drawings, it was later slightly modified to accept standard American components such as wheels and wheel-bearings, and was renamed the 57mm Gun M1.

By now the gun designers of each nation realised that a race was on between them and the opposing tank designers. Looking over the shoulders of their own tank prophets, they assumed that the enemy's forthcoming attraction would be at least as fast, thick and heavily armed, and begun to take the necessary steps to do something about it. Britain started in a small way by talking about an 8-pounder, whose performance would have been a marginal improvement on the 6-pounder. The pendulum then swung across and the 8-pounder was dumped in favour of a 4·5 inch gun firing a 55 pound projectile. The complete round of ammunition for such a weapon would have been almost beyond the lifting power

165 Germans examining 76·2mm Anti-tank guns.

166 German 75mm PAK 40 gun well sited to cover an approach road and provided with protection for the gunner.

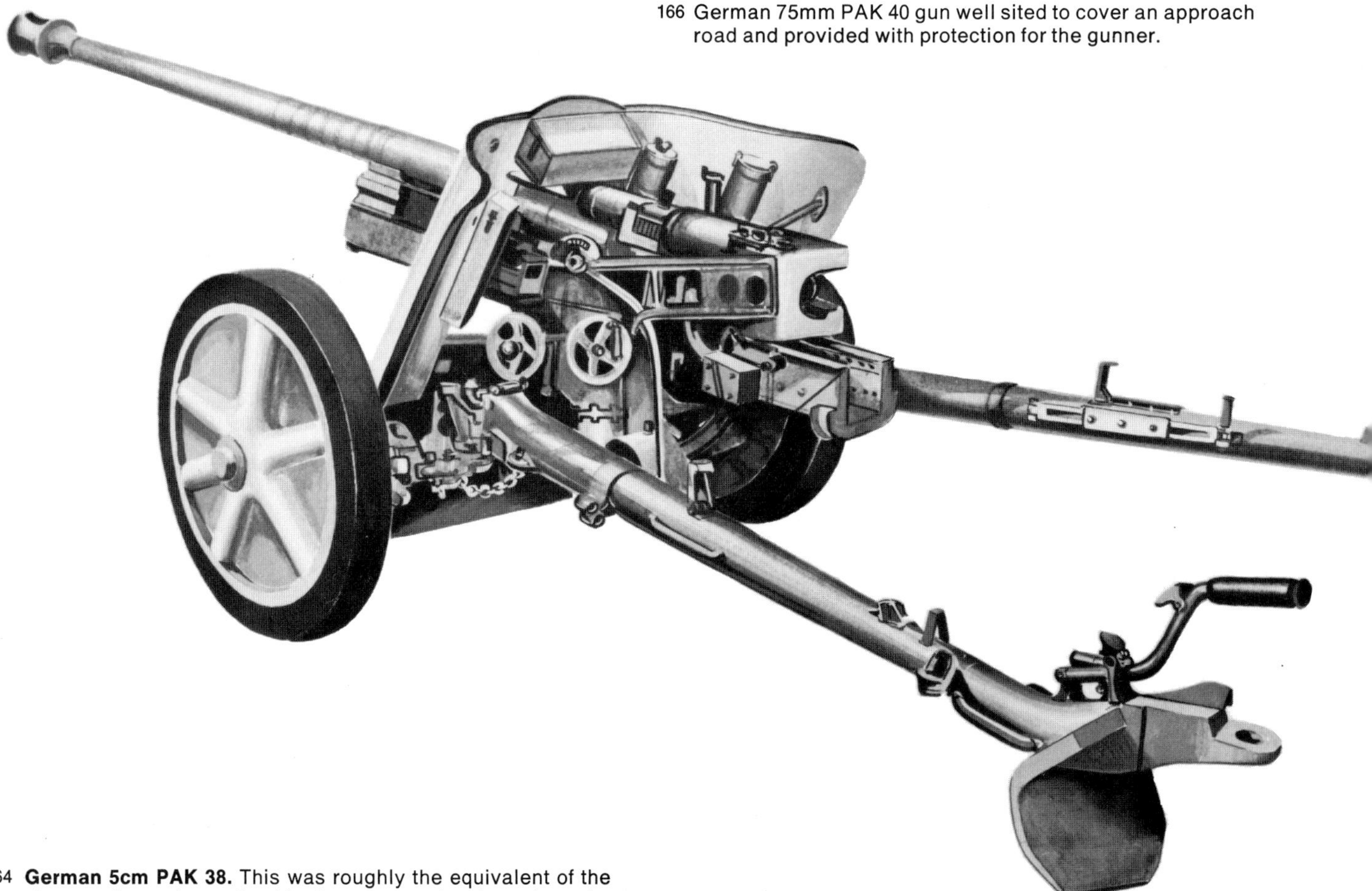

164 **German 5cm PAK 38.** This was roughly the equivalent of the six-pounder and remained in German service throughout the war.

25

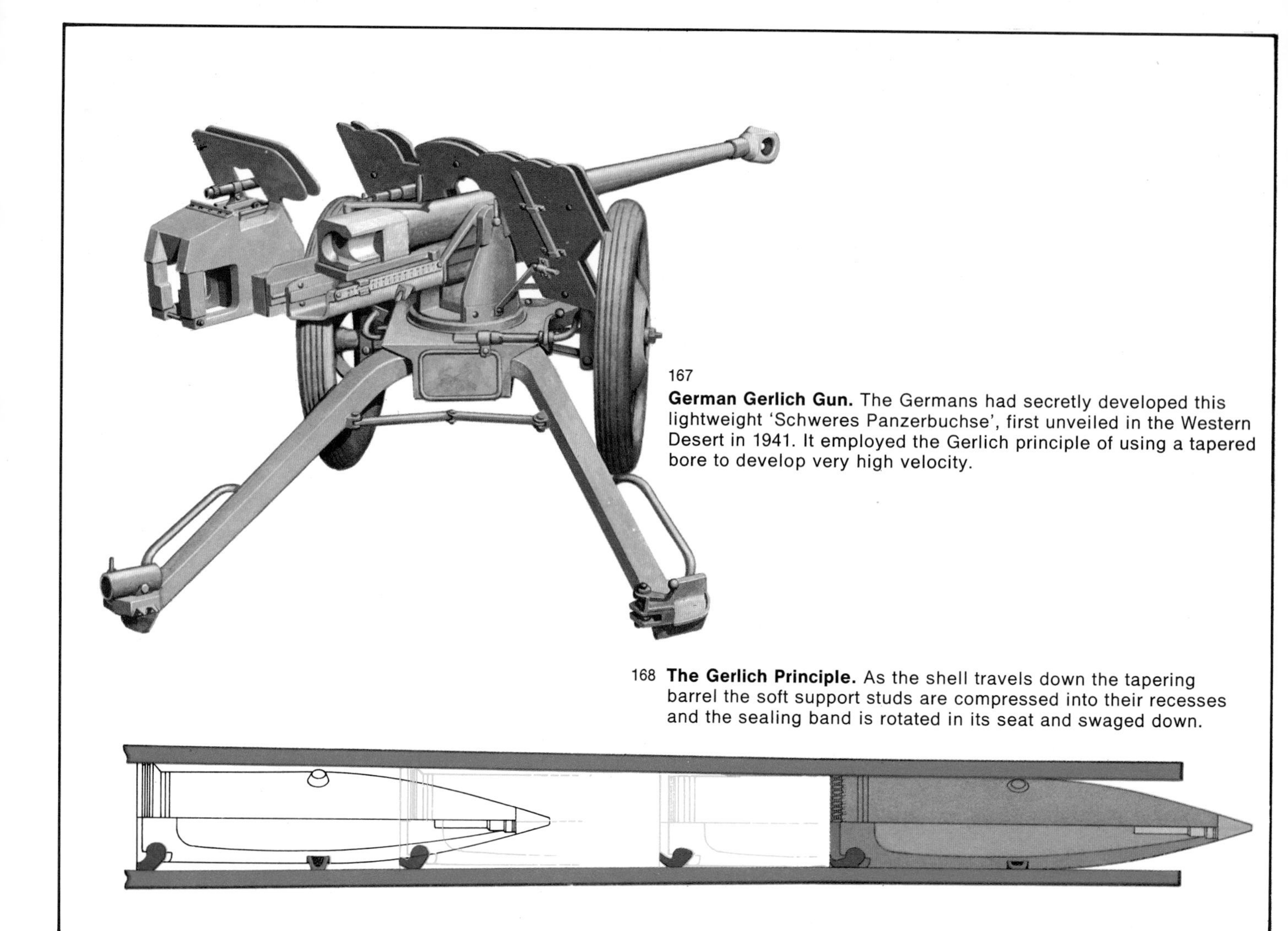

167
German Gerlich Gun. The Germans had secretly developed this lightweight 'Schweres Panzerbuchse', first unveiled in the Western Desert in 1941. It employed the Gerlich principle of using a tapered bore to develop very high velocity.

168 **The Gerlich Principle.** As the shell travels down the tapering barrel the soft support studs are compressed into their recesses and the sealing band is rotated in its seat and swaged down.

169 **Six-Pounder.** If you were in the British Army this was a six-pounder; if you were a GI it was a 57mm gun. Developed in Britain in 1938 but not issued until 1941, taken into use by the US Army too, it was a useful gun until the tanks began to get too thick.

of one man, and the problem of manhandling and concealing such a lump of ordnance was daunting. Fortunately sanity prevailed and the 55-pounder was dumped in its turn. Now the pendulum came to rest in mid-swing and three inches was selected as a convenient calibre. The designers went to work, the ballisticians predicted ample performance, and the project became known as the 17-pounder gun.

Germany, at the same time, was looking ahead in similar fashion; in 1940 a contract was put out for a heavier anti-tank gun, to be 75mm calibre. Rheinmetall who had produced a winner in the 5cm Panzerabwehrkanone (PAK) 38, simply scaled it up to produce the 75mm PAK 40 – so like its smaller brother that photographs without scaling features are easily confused (166). Krupp, on the other hand, went for the Gerlich principle once more and developed the 75mm PAK 41 (193), a taper-bore gun with an emergent calibre of 55mm. This was of unusual construction in that the shield formed part of the basic structure, the gun and cradle being trunnioned to it in a ball and socket joint and the trail legs being hinged from it. It was an ingenious design and deserved a better fate than was in store for it; indeed the US designers were so

170 **British Two-Pounder.** The British equivalent was this 40mm weapon, on a rather more ornate mounting which gave all-round traverse. It too was soon outclassed by the improvements in armour and tank guns.

171 **US 37mm Anti-Tank Gun.** The US Army's 1930s weapon was this 37mm model, very similar to that in service with the German Army at the time. Capable of dealing with the tanks of its day it was rapidly outclassed when heavier armour came on the scene.

taken by it that they adopted the same type of construction for a 90mm gun design which was developed later in the war (181).

In Britain, a rival to Gerlich had appeared. Before the German taper-bore gun was known, a Czech engineer working in Britain had proposed a squeeze-bore adapter to be fitted to the muzzle of the normal 2-pounder gun. With a special flanged shot, this would provide the benefits of a taper-bore in increasing the velocity of the anti-tank projectile without the complications of making a completely new gun, and allowing the basic gun still to be used with the normal projectiles (176). After a good deal of argument, the Director of Mechanisation managed to get the idea approved and a contract was given, at low priority, to develop the Janacek idea. While the work was going on, the German PzB 41 was brought into use, and a captured specimen was sent back to England, with a few rounds of ammunition, for testing. A routine check on velocity disclosed the astounding figure of 4050 feet per second, together with some impressive penetration figures. Obviously the squeeze-bore idea was a good one after all, and Janacek's adapter suddenly had its priority uprated. It would be nice, at this point, to relate a cliff-hanging tale of how the adapters were rushed to the front in time to have a decisive effect, but the truth of the matter is that when they were finally ready for issue the 2-pounder gun had been retired as an anti-tank weapon and replaced by the 6-pounder. Eventually the Littlejohn Adapter, as it became known (from the anglicisation of Janacek's name), was fitted to armoured car guns, and also made to fit the US 37mm armoured car guns, and on these vehicles it performed every bit as well as the inventor had forecast, remaining in service for many years.

Its only drawback was the necessity to attach the adapter to the gun muzzle before using the special flanged shot and removing it before firing the normal un-squeezeable AP shot and HE shells. The crews issued with these weapons found this to be a wearisome business, particularly in action, and soon some unknown genius observed that as the German Arrowhead shot (see below) was of similarly poor ballistic shape to the unsqueezed Littlejohn shot, but seemed to work just as well, then firing the Littlejohn shot without the adapter might very well work. So the adapters were left permanently off and the special shot fired from the un-tapering gun barrel. It saved a lot of work and the crews swore that it got the

172 Soviet anti-tank gunners in action, demonstrating the hazards of the trade.

same results, though doubtless Mr. Janacek would have been less than pleased had he known.

By now it was obvious that no matter how much bigger the tanks got, it was little use throwing steel shot at them; tungsten had to be adapted to all weapons to overcome the shatter effect at high velocity. But simply making a shot out of tungsten was no sort of solution; the material is so dense that if one were to make a 6-pounder shot out of tungsten it would weigh almost ten pounds, and thus the velocity would actually be less. The German solution was based on the Gerlich shot, without its flanges. They took a core of tungsten and then built it up to the necessary calibre by putting it in a steel or light alloy body. This made a projectile of the correct size, but whose weight was slightly less than the standard steel shot, thus giving a higher velocity. Britain adopted a similar projectile (174) and the Germans later came up with the "Arrowhead" shot (173) so called from its outline. This was a tungsten core with a steel surround in which the excess steel had been cut away, leaving only sufficient metal at full calibre to steady the shot in the barrel. All these projectiles had a good muzzle velocity, but since their ratio of weight to diameter was poor, they lacked "carrying power". This is a complex ballistic problem being reduced to simple terms; it can best be understood by comparing a golf ball with a ping-pong ball; if you launch both at the same velocity, which one will carry farther? The golf ball. Similarly a projectile with a poor weight/diameter ratio would soon lose velocity as it flew through the air, and these "composite rigid" projectiles soon lost their advantage, and at longer ranges were indeed worse than steel shot.

By the end of 1942 heavier weapons were on the way. The British 17-pounder had been completed but there was a delay in manufacturing the new split-trail carriage. Moreover the School of Artillery had been taken aback when they saw the pilot model. It was so heavy that it was barely possible for a full detachment to manhandle it over rough ground, and a good deal of "simplicate and add lightness" was called for.

Late in 1942 the first German 'Tiger' tanks had appeared in the pages of intelligence reports as a potential threat

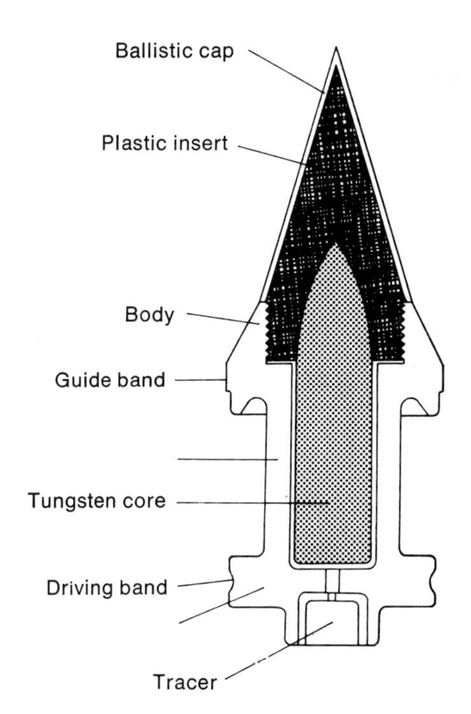

173
Arrowhead Shot. This German development, since copied by the Soviets, used a tungsten core and located it in a steel and plastic 'arrowhead' to reduce weight and so obtain high velocity.

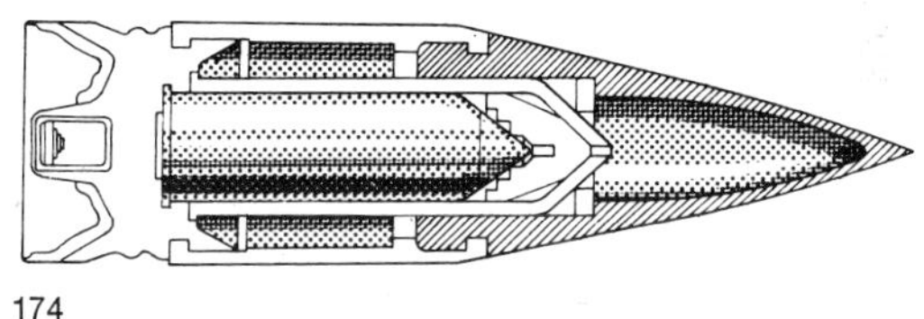

174
Composite Rigid Shot. Another method of achieving the same result, this British 6-pounder design has an external shape like the steel shot, but is air-spaced internally and has light alloy components around the tungsten core.

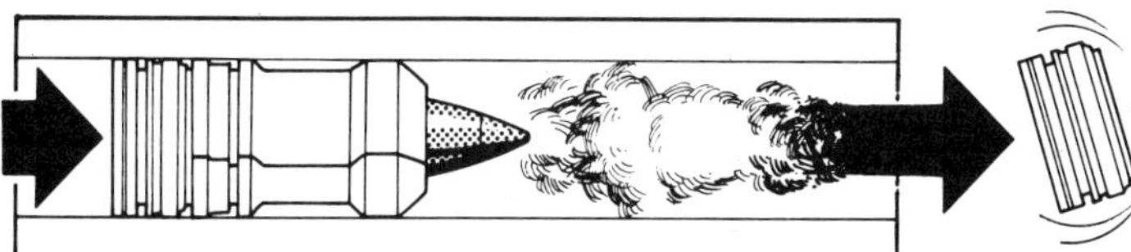

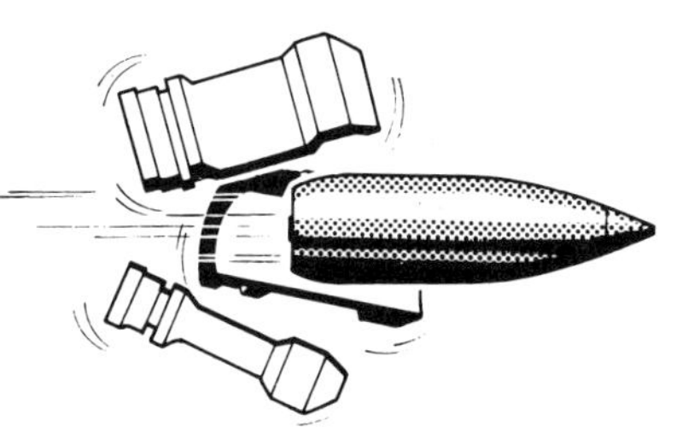

175 **Discarding Sabot Shot.** The tank-killer supreme. In the barrel it is a full calibre lightweight projectile, giving high velocity. On leaving the muzzle everything except the tungsten sub-projectile is discarded, giving the sub-projectile a good staying power to penetrate at long ranges.

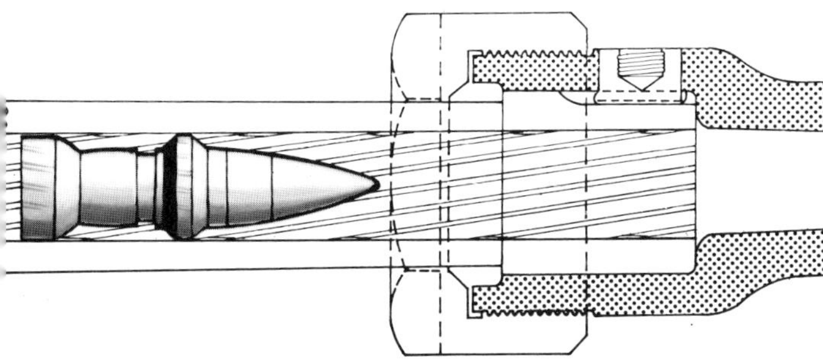

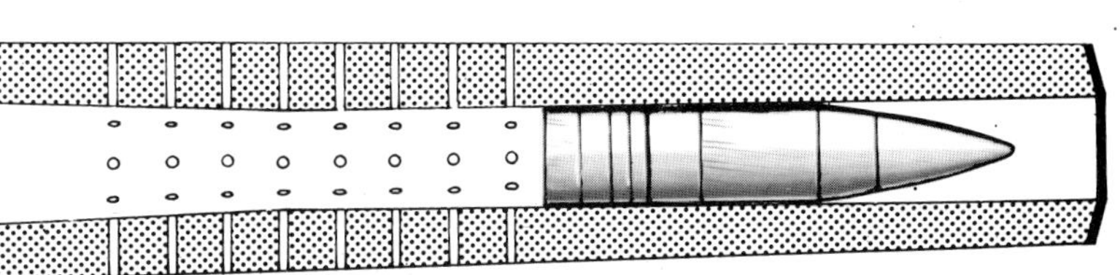

176 **Littlejohn.** A British method of adapting the Gerlich principle to a conventional gun is this coned adapter screwed to the end of the muzzle. Used with a special skirted projectile, it boosted velocity and penetrating power when fitted to British 2-pounder and US 37mm armoured car guns.

in North Africa, and soon after this notification a battery of the British 72nd Anti-Tank Regiment, armed with 6-pounders, managed to rout a tank attack near Robaa; when the smoke cleared three Tiger tanks were seen to have been stopped. (I am indebted to Sir Richard Duckworth, the then Battery Commander, for this information for this first Allied encounter with the Tiger seems to have escaped the history books.) Although the 6-pounder had won the day, it was a close-run thing, and as an emergency measure a hundred 17-pounder barrels were air-lifted to North Africa and hastily assembled to 25-pounder carriages to become the 17-pounder Mark 1 (177). They were then camouflaged to resemble 4.5 inch medium guns while in transit, and rushed to the forward troops in time to give the next Tiger attack a severe shock. It speaks volumes for the safety factor built into the ex-4.1 inch ex-25-pounder carriage design that it withstood the much greater recoil force of the 17-pounder without modification and without ill-effect.

The unfortunate German designers though were up against something more formidable than a mere tank: they were confronted in mid-1942 with a Fuhrerbefehl: "No more tungsten for ammunition – all supplies for machine tools." This was the death-knell of the Krupp 75mm PAK 41, since without tungsten there was no point in using it. At the same time Arrowhead shot was also finished for production, only the remaining stocks could be fired, and after that it was back to steel shot for the duration. After much argument, an exception was made for the 5cm PAK, since only a supply of Arrowhead could ensure stopping the Soviet T-34 tanks until there were enough 75mm PAK 40 guns to take over. This lasted until mid-1943, but after that tungsten-cored ammunition was a thing of the past for the German designers and gunners.

In England though there was not such a shortage, and the designers were working on the last and best of the composite projectiles, the AP Discarding Sabot shot. The tungsten core was enclosed in a steel sheath, which in turn was held in a full-calibre light alloy sheath. This was cunningly constructed so that it would fly to pieces when fired, separating from the core at the muzzle and leaving the sheathed core to fly to the target. Thus inside the gun there was ample base area for the large cartridge to push against, but in flight there was only the small sub-projectile, with its excellent weight/diameter ratio and carrying power, screaming through the air at 3000 and more feet per second (175).

This APDS was first issued for the 6-pounder in early 1944, followed by issues for the 17-pounder in September 1944. 17-pounder cores were also supplied to the USA for them to manufacture APDS for their 76mm and 90mm tank and anti-tank guns, though they did not get this type of ammunition into service till some years after the war, preferring composite rigid projectiles. APDS has since been

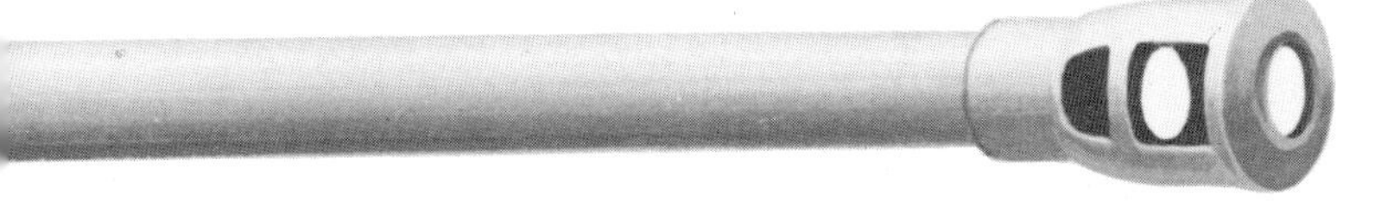

177

The Seventeen/Twenty-Five Pounder. When the 17-pounder was put into production, carriages were slower off the line than guns, and a number of barrels were mounted on 25-pounder carriages as an interim measure. It says much for the robustness of the 25-pounder carriage that it managed to stand the strain.

adapted to every British high-velocity anti-tank gun to the present day.

The continuing story of the wartime race for supremacy has rather caused us to leap ahead and ignore one or two other wartime fields of anti-tank activity, which it may now be as well to look at.

Britain in 1940 was far from well off for anti-tank weapons; most of the 2-pounders were the far side of the Channel, and the unfortunate 6-pounder still hadn't found itself a factory. Many and varied were the ideas put forward by various inventors in the hopes of finding a cheap and cheerful anti-tank weapon, though most of them were infantry hand-held weapons. One attempt to provide something heavier was to take fifty old 3 inch 20 cwt. anti-aircraft guns and mount them into Churchill tanks, manufacturing a suitable AP shot to suit (230). Later, when 17-pounder carriage production began to outrun gun production, another fifty were put onto 17-pounder carriages for home defence while the 17-pounders went to the field armies. I doubt very much whether any were fired in anger, but they would have been a serviceable enough weapon had the need arisen.

Many and outlandish were the ideas produced by and for the Home Guard, Britain's civilian volunteer defenders. The prime requisite here, assuming the design worked, was that it should be cheap and easy to make. Many

178 A rare photograph showing "Barndoor" the 88mm PAK 43/41, in action in Russia. The wheel and trail leg slowly subsiding into the mud illustrate the defect of heavy anti-tank equipments.

182

254mm case hardened armour converted for air conditioning by a 225mm shell. Photo Bofors A.G.

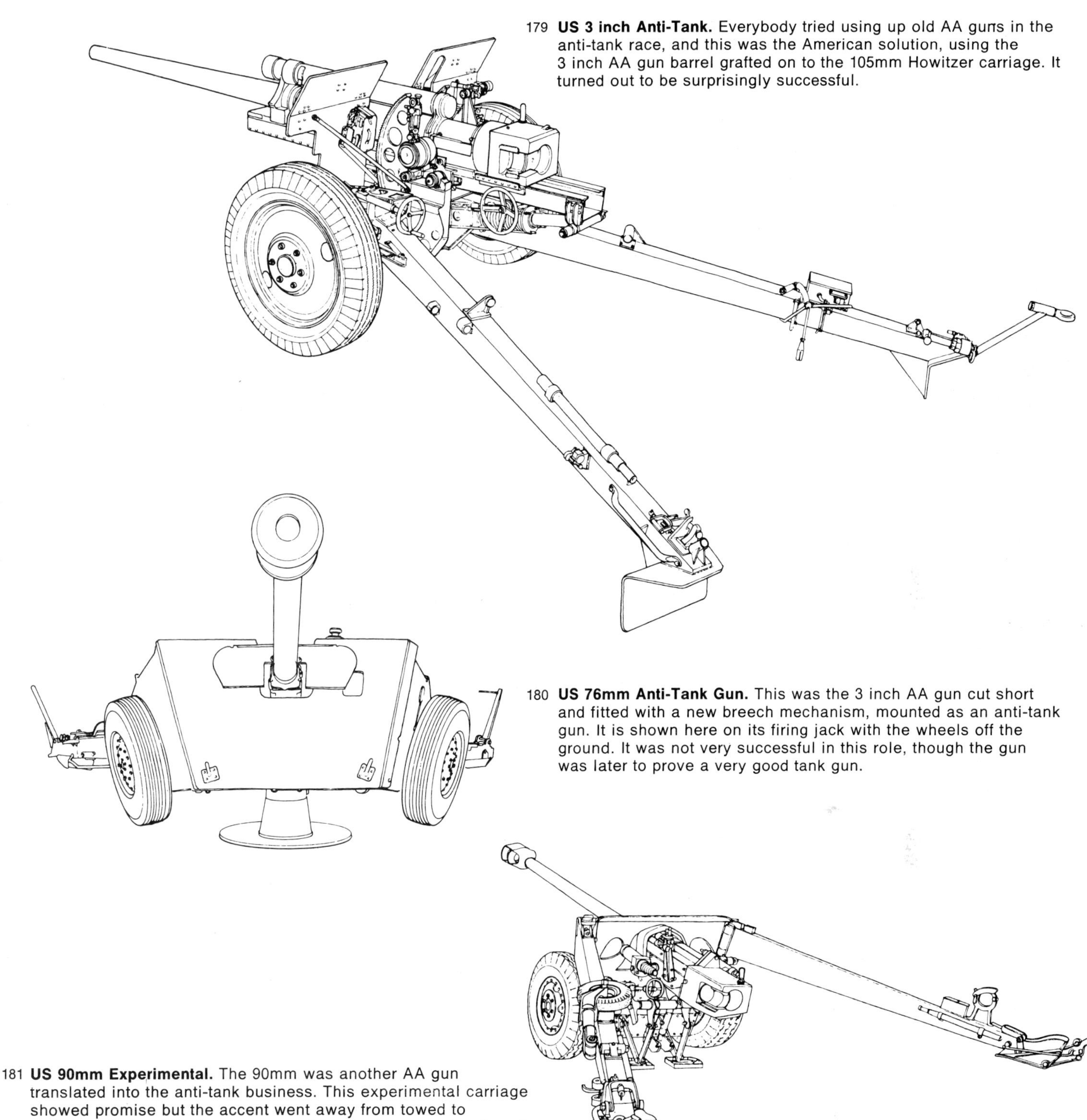

179 **US 3 inch Anti-Tank.** Everybody tried using up old AA guns in the anti-tank race, and this was the American solution, using the 3 inch AA gun barrel grafted on to the 105mm Howitzer carriage. It turned out to be surprisingly successful.

180 **US 76mm Anti-Tank Gun.** This was the 3 inch AA gun cut short and fitted with a new breech mechanism, mounted as an anti-tank gun. It is shown here on its firing jack with the wheels off the ground. It was not very successful in this role, though the gun was later to prove a very good tank gun.

181 **US 90mm Experimental.** The 90mm was another AA gun translated into the anti-tank business. This experimental carriage showed promise but the accent went away from towed to self-propelled anti-tank weapons and the design was abandoned in 1945.

Home Guard units, especially in industrial areas had manufactured simple mortars and bombs to suit, but an anti-tank weapon was a slightly more difficult proposition. Eventually three officially approved weapons were issued, the Northover Projector, the Blacker Bombard, and the Smith Gun. The Northover was a two-inch calibre smooth-bore barrel mounted without benefit of recoil system onto a simple tripod; the breech mechanism was positively agricultural. The standard projectiles were the No. 36M hand grenade, the No. 68 hollow charge rifle grenade, and the No. 76 Self-igniting Phosphorus Grenade. The 36M (better known as the Mills Bomb) was a lethal anti-personnel projectile, and the other two were quite satisfactory against armour, but the maximum range with any certainty of a hit was not much more than a hundred yards and the muzzle velocity was so low that attacking a moving target was more like a game of chance.

The Blacker Bombard was a spigot mortar, a species of

DUMMY
57R
HEAT
16824
L31 LT

183 **Typical Cartridges.** *From left to right:*

A percussion primer

A sectioned 25-pounder cartridge, showing the colour-coded charge bags

A wrapped steel case for the US 105mm Howitzer

A bagged charge for the British 5.5 inch gun, showing the igniter pad at the rear end, and with two tubes in the foreground

184 **Recoilless Gun Ammunition.** *From left to right:*

British 3.45 inch (Burney) wallbuster shell with its perforated brass cartridge case attached

US 57mm (Kromuskit) round, with a hollow charge shell and with the perforated case made of steel

German 105mm blow-out-base cartridge case showing the recess for the side-mounted primer

185 **Fuzes.** *From left to right:*

British direct action Fuze No. 49, designed for attacking Zeppelin fabric

British base fuze No. 12, for armour-piercing naval shells

Wartime proximity fuze for British 25-pounder gun

Postwar proximity fuze for US field guns

British base fuze for early wall-buster shells

British percussion fuze No. 117, for high explosive shells

British mechanical time fuze No. 213 for high explosive airbursts

186 **Carrier Shells.** *From left to right:*

US 75mm White Phosphorus smoke shell, sectioned to show the central explosive burster

French 75mm White Phosphorus smoke shell with extended fuze and short burster, a First World War design

Swedish Carl Gustav 84mm Illuminating shell sectioned to show the star unit and the folded parachute beneath

British 25-pounder base ejection smoke shell, showing the three smoke canisters, ejection charge and time fuze

187 **Anti-Aircraft Projectiles.** *Left to right:*

German 5cm Flak HE shell

British 6-pounder experimental AA gun HE shell

Swedish 40mm Bofors HE shell

British round for Bofors 40mm L/60 (World War Two) gun

British round for Bofors 40mm L/70 (Post-war) gun

188 **More Anti-Aircraft Projectiles.** *Left to right:*

German 88mm Flak HE with grooves in the body to improve fragmentation

British 3.7 inch Mark 6 Gun HE shell, showing the driving and steadying bands associated with the Probert system of rifling

German 37mm 'Flaming Onion' incendiary shell

189 **British Anti-Tank Discarding Sabot Projectiles.** *From left to right:*

6-pounder, ca 1944

20-pounder 'Pot' sabot, a Canadian design

17-pounder 'Petal' sabot

105mm 'Pot-Petal' sabot, currently used by most NATO forces

120mm Chieftain tank gun practice shot

190 **Anti-Tank Rounds.** *From left to right:*

Soviet 76mm Arrowhead shot

British 105mm APDS

US 37mm Armour-piercing shot

US 90mm armour-piercing shell, showing the structure of the piercing and ballistic caps, the small explosive filling and the base fuze

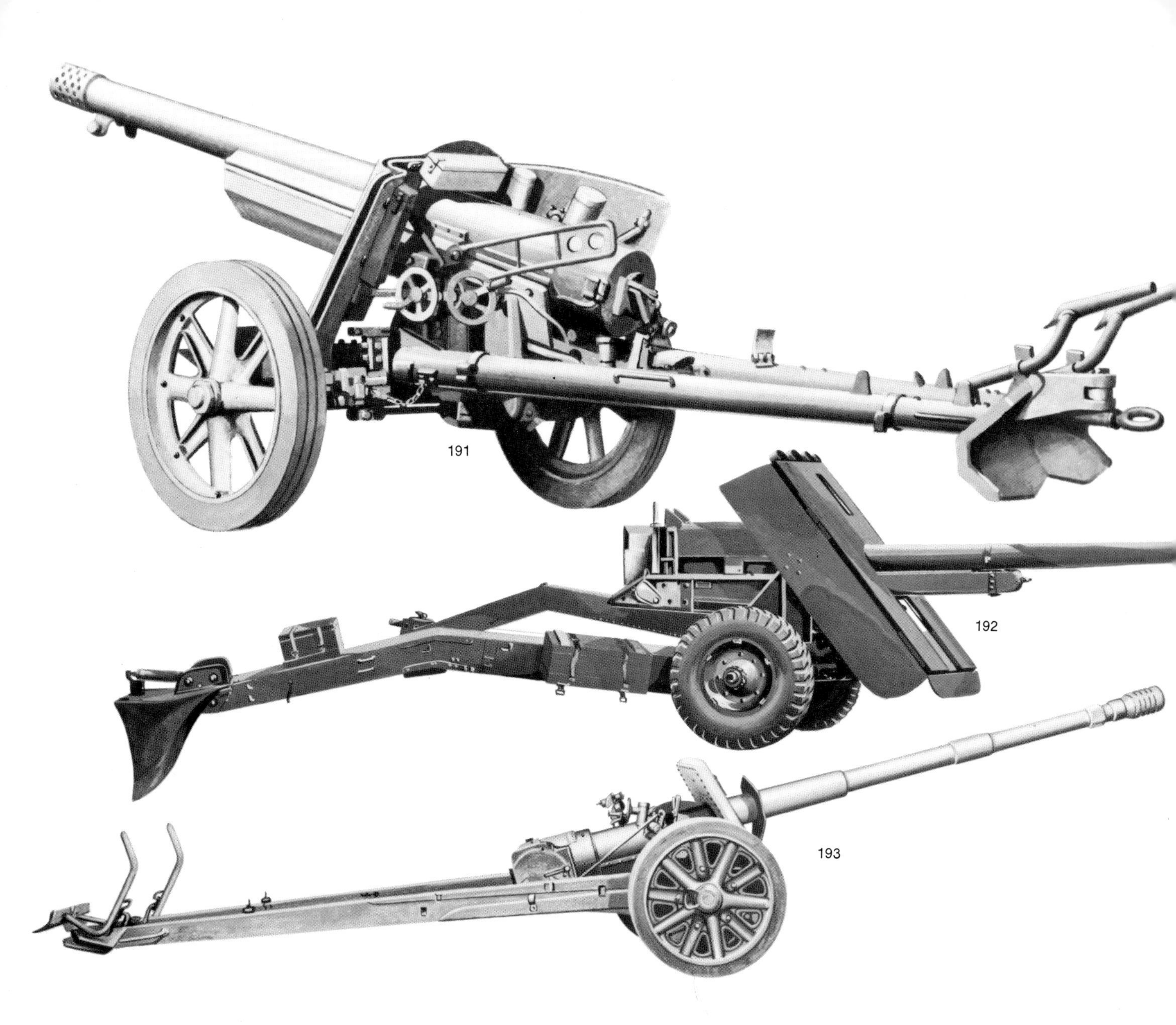

weapon which Colonel Blacker had been assiduously designing, promoting and patenting throughout the 1930s. His principle was simple; do away with the gun barrel and you do away with what is probably the most expensive and time-consuming manufacturing process in the entire weapon. So he made his projectiles in the form of bombs with long hollow tail booms carrying stabilising fins, and positioned the propelling charge at the front of this hollow tail. To fire, the bomb was supported by some suitable frame, and a heavy steel rod (the 'spigot') driven by a powerful spring, was forced into the tail boom. It struck and ignited the charge and the explosion then blew the bomb off the spigot, contact between tube and spigot giving initial direction to the bomb. At the same time, the force of the explosion would also blow the spigot back into its housing, recompressing its spring, where it could be held cocked ready for the next round. Mounted on a low ground-hugging tripod the Bombard was easy to conceal and carried a powerful charge of explosive in its 20-pound bomb.

The most ambitious weapon was the 3-inch calibre smooth-bore Smith Gun. This was the design of a Mr. Smith, an engineer with the Tri-ang Toy Company, and its most conspicuous feature was its carriage. The wheels were conical, and arranged so that once in position the whole weapon could be tipped over to sit on one wheel which had its flat side outwards, and this acted as a pivot mounting. The other wheel had its conical section outwards and this now gave overhead protection, while a simple shield filled the space between and practically provided a turret for the crew. It fired a simple cast-iron shell fitted with a modified hand grenade fuze and filled with high explosive, and for anti-tank work was provided with a fin-stabilised hollow charge bomb which was little more than an overgrown No. 68 rifle grenade; for all its oddities, it was an effective weapon, and after the Home Guard had their allocation, they were issued to Regular units for the local defence of airfields in Britain.

The U.S. Army, while never having to resort to anything as stark or primitive as the Northover or Smith guns, were

191 **German 75mm PAK 97/38.** When the German Army over-ran France in 1940 they captured hundreds of French 75mm guns. These were later converted into extempore anti-tank guns and used extensively on the Russian Front.

192 **Seventeen Pounder.** The British 17-pounder (3 inch calibre) was one of the best anti-tank guns to come out of the Second World War. When discarding sabot shot had been produced for it, it could defeat anything.

193 **German 75mm PAK 41.** Krupp's entry into the 75mm anti-tank field, this was a brilliant design. The carriage structure was novel and the gun used a taper bore. It could take on and defeat any tank in the world until the supply of tungsten for the special ammunition ran out.

194 **Soviet 76.2mm Gun.** In the early weeks of the invasion of Russia hundreds of Soviet field guns were captured by Germany. These were converted to take a German cartridge case and became very efficient anti-tank guns.

not above a little discreet redeployment, and took their redundant 3-inch Anti-aircraft gun M1917 and fitted it to surplus 105mm carriages to form the 3 inch Anti-tank Gun M5 (179). It was later fitted into a much-modified Sherman tank chassis to become the 3 inch Gun Motor Carriage M10, but that story belongs elsewhere. Finding that 3 inches seemed to be about right for an anti-tank gun, the next step was to improve the existing model; recontouring the chamber and improving the projectile turned it into the 76mm gun, but after fooling around with some field carriages, this eventually became a tank gun. At the same time, since ex-AA weapons seemed to be the in thing, the 90mm AA gun was taken as the starting point for the next generation of anti-tank guns. As already mentioned, the designers threw tradition out of the window with this one, and produced some way-out designs, though none of them ever reached production.

In Britain, whether the American 3 inch to 76mm transformation had been observed or not, a similar trick was done by removing some of the barrel of the 3 inch AA gun, fitting 17-pounder projectiles to the cartridge, and issuing the result to tanks as the 77mm gun.

The Germans, past masters at the art of cobbling guns together from unlikely components, found themselves short of anti-tank weapons in the early part of 1942 when the Russians began to collect their wits and put up a stiffer resistance to the German invasion. Their reaction was to round up all the French 75mm M1897s they could find, strip off the barrels and fit them with muzzle brakes, and then drop them onto 5cm PAK 40 carriages. To cope with armour though, since the cartridge was a relatively low-velocity model and insufficiently powerful to use with solid shot, some new ammunition had to be developed (191).

Their second extemporisation was to take the many captured Soviet 76mm field guns, which were good sound weapons, ream out the chambers to take a heavier cartridge, provide an AP shell or shot, and send them back to the front as the 76·2mm PAK 36(r) (r for Rusland). This gun performed very well, and the ammunition which the Germans developed was so effective that it formed the basis of many Russian post-war designs (194).

In Britain, with the 17-pounder coming along, thoughts now turned to what to follow it up with. At least a 25 per cent improvement in ballistic performance was called for, and whether the German use of their 88mm AA gun as an anti-tank, or the US 90mm conversion, or the British 3 inch conversion started a train of thought will never be entirely clear, but the upshot of it all was the proposal to take the barrel of the 3·7 inch AA gun and, with a little sawing here and there, graft it onto a suitable carriage and make it into an anti-tank gun.

This project became known as the 32-pounder, and designs of towed and self-propelled carriage were begun, as well as development of suitable ammunition. The towed carriage (205) was a heavy and complicated piece

195
The Blacker Bombard manned by a Home Guard Squad.

of machinery incorporating a hydraulic system for retracting the gun for travelling, a split trail and an enveloping shield. It was test fired in 1945 and was undoubtedly among the most powerful and formidable tank killers ever built, but as a practical field weapon it was far too cumbersome, and only two were ever built. The self-propelled version, named "Tortoise", was little more successful, being of vast bulk and weight and having the gun mounted frontally so as to give limited traverse – in many respects the British equivalent of the German "Ferdinand". Six were built but they were not completed until well after the war had ended, and after being used for a variety of trials four were scrapped and two relegated to museums.

One famous weapon which has hardly been mentioned yet was the German 88mm gun. This had its inception in the 1930s when Krupp designers were working with Bofors of Sweden. The terms of the Versailles Treaty were such that there was little employment for gun designers in

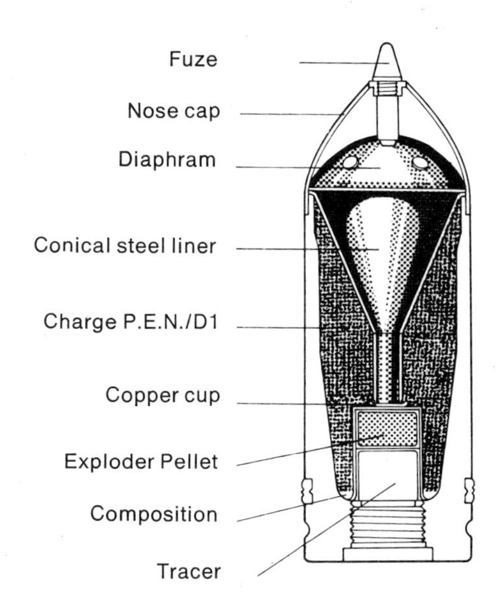

196
Hollow Charge Shell. A typical early design utilising the hollow charge principle is this British 95mm model. The impact fuze is designed to fire a miniature hollow charge jet back into the shell to initiate the explosive filling.

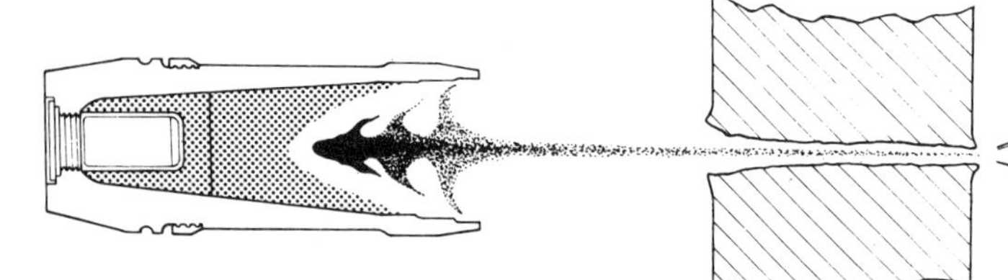

197
The Hollow Charge. For a low velocity gun to have an anti-tank capability it must employ the explosive force contained in the shell instead of relying on high velocity. One way is to utilise the hollow charge principle in which the cone of explosive focusses the detonation into a jet of hot gas which forces its way through the target plate.

Germany, and they were farmed out to foreign companies to keep their eye in. In the early 1930s they returned to Essen bearing the drawings of the 88mm gun. Pilot models were built, and due to the thorough thinking-out enforced by the Swedish exile, all the designers' hopes were realised. It was immediately accepted by the Luftwaffe, and as an anti-aircraft gun we have already met it. Early models were taken to Spain with the Condor Legion, and there its potentialities as an anti-tank gun were realised. It was provided with suitable sights, and armour-piercing projectiles were produced. As an anti-tank gun it was rarely, if ever, used in the Polish and French campaigns, partly because the tanks of the time could be effectively dealt with without calling on the 88 and partly because it was a gun best suited to defensive work. So it was not until the later campaigns in Africa that the 88 became well-known as a tank destroyer. In view of their success, Krupp's were now called upon to produce a proper anti-tank weapon, rather than a worked-over anti-aircraft gun, and they began a grandiose scheme for an integrated family of AA, A/Tank, U-boat and tank 88mm guns which would all fire the same ammunition and use common parts, but slowly the accent moved away from this ideal system and concentrated on the tank and anti-tank weapons. The anti-tank gun eventually produced was the PAK 43, loosely based on the original AA gun and having a similar type of mobile mounting, but with more powerful ballistics and, since it was only required to elevate to 20 degrees, a lower and more easily concealed outline. Gun production outstripped mounting production (not an unusual state of affairs) and the first guns were adapted to a two-wheeled carriage hastily thrown together from a selection of parts they had handy – the wheels of the 15cm howitzer, trail legs from the 10·5cm gun and so forth, and this was rushed into action on the Eastern Front, known as the PAK 43/41. Cumbersome and awkward to handle – it was nicknamed "Barndoor" – it was nevertheless an effective gun and served well until the proper PAK 43 came into use.

Before the PAK 43 was operational though, it was obvious that something heavier would soon be wanted. The German Army had been impressed by the Russian 122mm guns and asked both Krupp and Rheinmettal to produce an anti-tank gun of similar calibre. Both elected to use 128mm and produced very similar prototypes (200). These PAK 44's had cruciform carriages with low silhouettes and promised to be formidable weapons, but they came too late. Before they could be brought to production the war was over. As usual, barrel and breech production was well ahead of carriage production and as a stop-gap measure and to obtain firing data, numbers of barrels were mounted on a variety of carriages, notably the ex-Russian 152mm gun (202) and the ex-French 155mm gun carriages. Due to the difficulty of matching the recoil systems to the new and powerful gun, these adaptations were not particularly successful.

Earlier, we briefly discussed the German 75mm 97/38, the modified French 75mm barrel on a PAK 40 carriage, and pointed out that the velocity was too low to expect any worthwhile results with piercing projectiles. Similar problems had arisen in other countries. Although anti-tank guns are deployed to catch tanks, some will inevitably dodge the guns and arrive in front of some non-anti-tank weapon. The Indian Army wanted an anti-tank projectile for the 3·7 inch Pack howitzer; the US Army wanted one for their 105mm and 75mm howitzers. In every case a piercing projectile was out of the question. Consequently a new type of shell was developed in which explosive force replaced kinetic energy. This was the hollow charge shell.

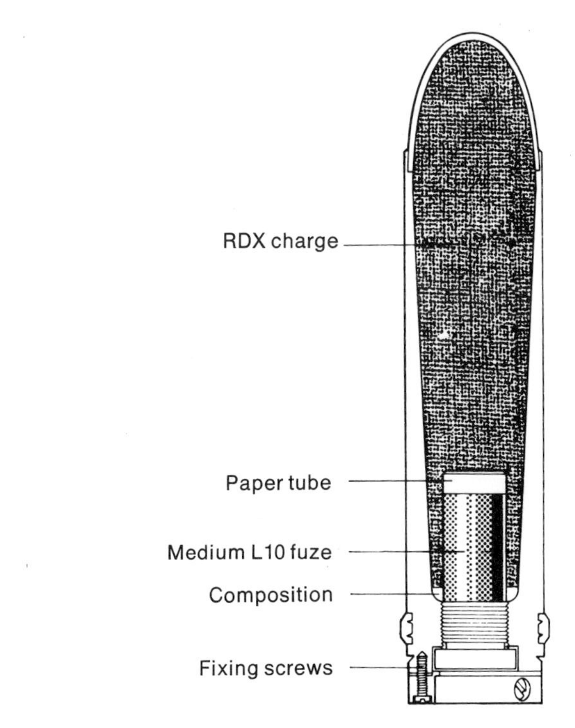

199

Squash-Head Shell. The Squash-head shell uses a thin body and a filling of soft explosive together with a base fuze. On impact the thin nose peels away to allow the explosive to squash on to the armour.

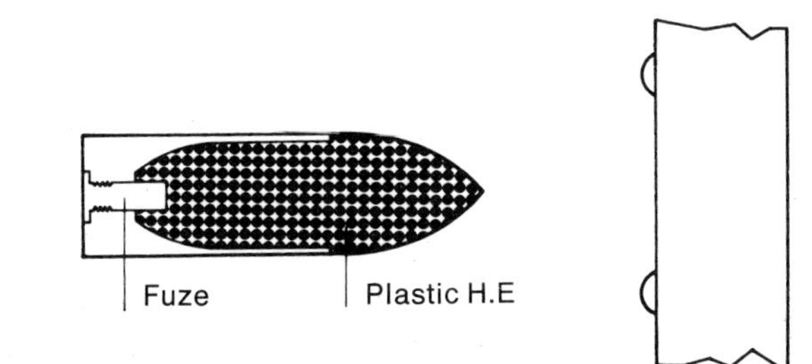

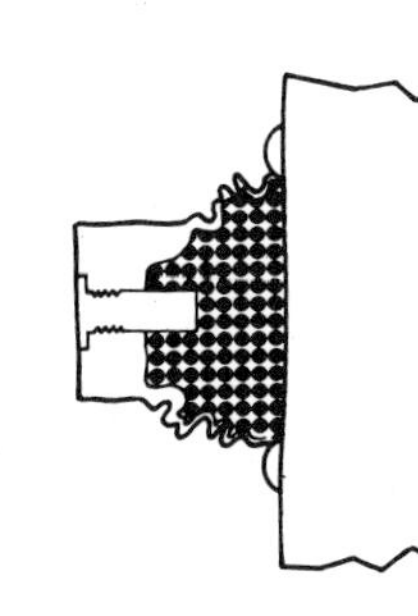

198

The Squash-Head Method. Another low velocity solution is to squash a pancake of explosive on to the face of the armour and detonate it to drive a 'scab' off the interior face at high speed.

The hollow charge effect was not new; the American engineer Monroe had described it in the 1880s after observing that the letters USN incised in slabs of guncotton reproduced themselves in the surface of a slab of steel after the guncotton had been detonated in contact with it. After a series of further experiments he showed that preparing a hollow in the face of the explosive enhanced its cutting or penetrating effect, but it remained more or less a scientific parlor trick (the 'Monroe Effect') for several years. During the First World War some attempts were made to try and find a warlike application, without much success, and then in the early 1920s Neumann, a German experimenter, found that lining the hollow with metal gave a greater penetrative effect, and also that 'standing off' the charge, so as to leave a space between it and the target, improved performance even more (197).

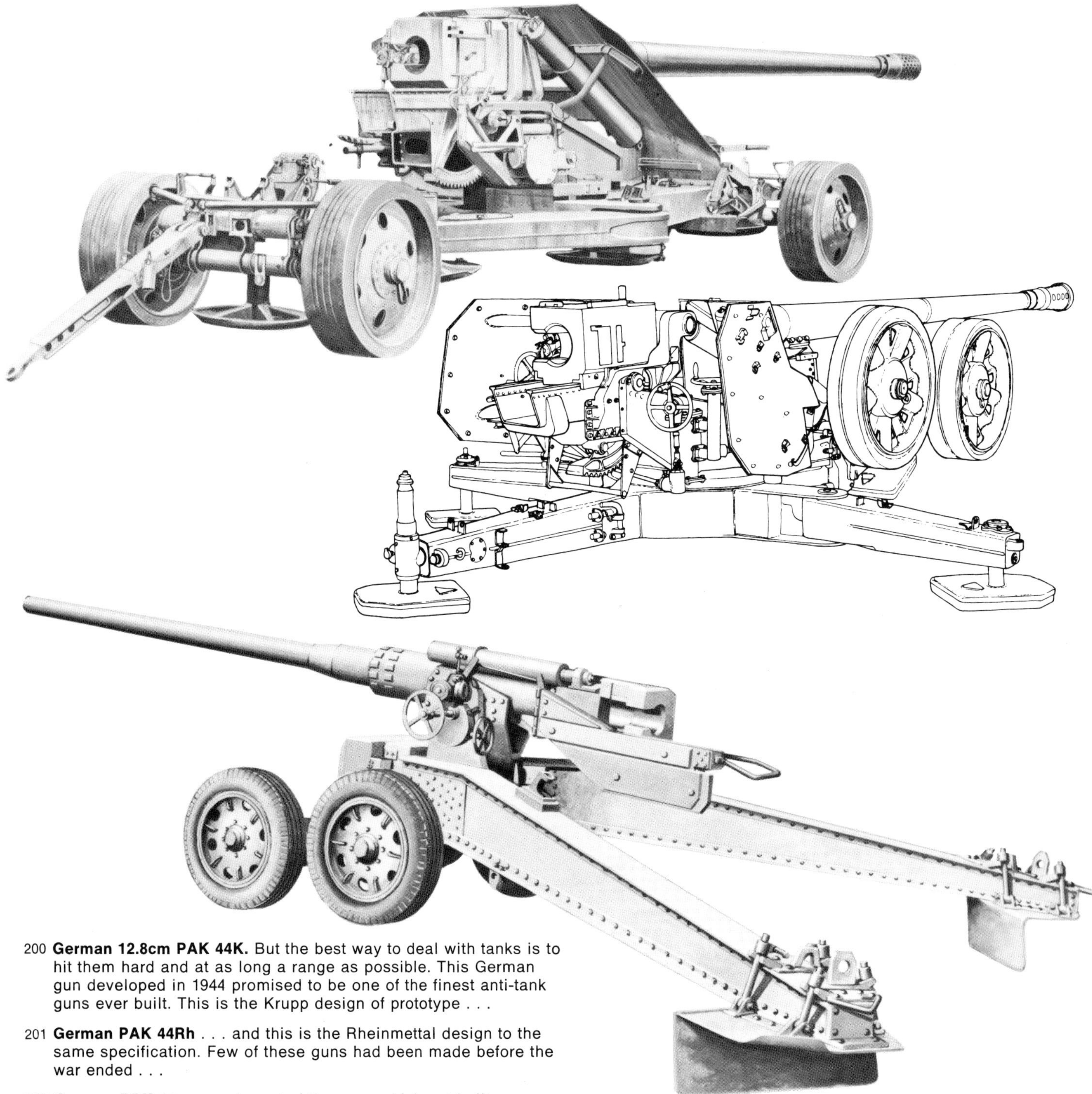

200 **German 12.8cm PAK 44K.** But the best way to deal with tanks is to hit them hard and at as long a range as possible. This German gun developed in 1944 promised to be one of the finest anti-tank guns ever built. This is the Krupp design of prototype . . .

201 **German PAK 44Rh** . . . and this is the Rheinmettal design to the same specification. Few of these guns had been made before the war ended . . .

202 **German PAK 44** . . . and most of the guns which got built were mounted, for trials, on Russian 122mm carriages.

It was not until 1938 that serious work began on trying to make a piercing weapon, and by 1940 the British Army were equipped with a hollow charge rifle grenade. The principle of operation was imperfectly understood at the time and designs were largely empirical but by mid-1940 hollow charge was the coming thing, and it was being considered for everything from grenades to ship torpedoes. But gradually some disadvantages began to appear.

The operation of the hollow charge is complex, and was not fully understood until high-speed photography, spark photography and X-rays had been adapted to examine the action, but it can be described in a simple way; the explosive charge is initiated at a point remote from the

203
A Bofors 37mm anti-tank gun being fired on an intrumented range to determine velocity. Photo: Bofors A.G.

204
Soviet 100mm Anti-Tank Gun. During the Second World War the Soviet anti-tank guns were an unimpressive lot, but the lesson must have got home, for in post-war years this powerful weapon was introduced.

205 **British 32-Pounder.** The British developed their 3.7 inch anti-aircraft gun into a monster anti-tank gun which, fortunately for the gunners who might have had to push it, arrived too late to take part in the war.

206
US 75mm Pack Howitzer This low velocity howitzer, developed for mule transport and widely adopted for airborne use, is typical of the weapons which need hollow charge or squashhead shells to give them a chance against tanks.

207 **Soviet 57mm Anti-Tank.** Although the British and US Armies had abandoned the 57mm as being too small, the Soviets produced this weapon in 1943 and kept it in service until well after the war.

208 **Japanese 47mm Anti-Tank Gun.** Japan's anti-tank armoury was relatively low-powered, demonstrating the pitfall of basing your anti-tank guns on your own ideas of tank design.

209 **German 75mm PAK 40.** Virtually a scale-up of the successful PAK 38 design, this was another stalwart which stayed in service throughout the war.

shaped section so that the detonating wave passes through the mass of explosive and then reaches the shaped liner in a symmetrical fashion. As the wave passes over the liner the metal is deformed and flows, performing as a liquid due to the intense pressure exerted by the wave. The liner and the explosive wave are 'focussed' into a fine jet aligned with the axis of the hollow, this jet being comprised of finely-divided metal and hot gas and travelling at speeds in excess of 20,000 feet a second. Due to the 'stand-off' or distance of the charge from the target, the tip of the jet actually accelerates and strikes the target armour with intensely concentrated force, forcing the metal of the target aside and boring a hole.

The principal disadvantage which made its appearance as more hollow charge weapons were designed was that where the projectile was spinning, the jet was diffused due to centrifugal force and projectile yaw, so that instead of concentrating its effort into a small area it spread out and thus reduced the impact and depth of penetration.

In view of this, hollow charge came to be used almost exclusively on fin-stabilised projectiles such as the Bazooka, the Panzerfaust, and a variety of rifle-launched grenades. This meant that the infantryman now carried a very powerful punch indeed, being capable of launching, from his shoulder, a projectile which could hole the thickest tank.

However a large number of guns were provided with hollow charge; it may not have been working at optimum performance, but even with the defect of spin it was still producing a better answer than a low-velocity shot would have done; the German PAK 97/38, the US 75mm and 105mm howitzers, the British 3·7 pack howitzer and 95mm Infantry gun all were provided with hollow charge; almost every German field gun had one or more designs of hollow charge shell, and the Russians copied German designs. Towards the end of the war the Germans began to take steps to combat the spin effect by designing shells with loose driving bands and spring-out fins, so that they could be fired from rifled guns without taking up much spin, and then would travel through the air fin-stabilised. It is of interest to see that most of these ideas were revived in later years by various nations and the US, for example, use a fin-stabilised round for their 106mm gun which is basically derived from German designs (278).

Britain was the first nation to have a hollow charge weapon in service, and they were the first to perceive the spin defect. Fortunately, at the time, another projectile had been developed which was immediately adopted for gun use to replace the hollow charge; this was the Wallbuster shell, later renamed the Squash-Head shell, known in the USA as HE Plastic. This was the invention of Sir Denis Burney, a noted British engineer, who was very much involved with the development of recoilless guns, which we shall discuss in a moment. In the development of his gun, he also began to develop a special shell to take advantage of certain characteristics of the gun. It was a high-capacity thin-walled shell, in which the HE charge was enclosed in a wire mesh bag with a base fuze attached (266). On striking the tank, the shell body would peel away like a banana, leaving the mesh bag to crush up tightly against the armour plate. When this crush-up had taken effect, and the explosive was tightly spread on the plate, the fuze detonated the charge and set up a shock wave inside the plate. This overstrained the plate and blew a large scab from the inside face into the tank at high velocity, demolishing all in its path. It was originally conceived as an anti-concrete weapon, and at its first demonstration in 1943 a 7·2 inch Wallbuster was fired against a five-foot thick wall. It blew lumps off the far side for sixty yards, and this result was so startling that its application to armour was soon tried, with equal success. For the rest of the war this type of shell was confined to use in recoilless weapons, since its design was too light for a conventional gun pressure. But it was redesigned to do away with the wire mesh bag, the filling was made of plastic explosive, and shortly after the war it was produced in a form suited to use in any kind of gun. It is still widely used today and is still a formidable tank killer (199).

Self-propelled guns

210 An early attempt at self propulsion. The Davidson steam car of 1900 mounted a Colt machine gun. Built by the Northwestern Military and Naval Academy, it failed to impress the military of the day.

The greatest wartime sphere of development of conventional artillery was the rapid rise of the self-propelled gun. Indeed this burgeoning had led most people to the assumption that the SP gun was a wartime invention: so it was, but it was the first World War, and not the Second, which saw the beginning of the SP gun. As soon as the tank had made its mark in 1916 there were suggestions to endow artillery with the same degree of mobility. Some of the early tank designs – the more grandiose 'landship' ideas particularly – envisaged the tank as little more than a protected gun platform but the limitations of size and power soon cut these ideas down and small direct-fire weapons became the tank's standard armaments. Early French tank designs used the ubiquitous 75mm M1897 as their main armament, though these were purely direct-fire weapons in this application and there was no possibility of deploying the tanks as indirect-fire weapons. In the long run the unreliability of these early models led the French Army to concentrate on smaller types such as the Renault two-man tank, armed with nothing heavier than a machine gun.

The British, as befitted the innovators of the tank, made the first tentative steps towards the SP gun, with the introduction of the 60 pounder gun carrier. This was a highly modified tank chassis onto which a 60 pounder gun was run and the wheels removed. While it probably could have been fired from this position, the idea was never seriously explored, the purpose behind the vehicle being nothing more than to transport the 60 pounder across the shell torn battlefields. Quite frankly the gunners didn't think much of the idea; it took a good deal of heaving and sweating to get the 60 pounder onto the carrier and remove its wheels, the loaded carrier could move no faster than a slow walking pace, and having arrived at the gun position the gun had to be fitted with its wheels and heaved off the carrier. With a little care in map reading, the road network in Flanders was sufficient to allow wheeled guns to be towed into practically any position required of them, and after some desultory experiments the gun carriers were withdrawn, refitted as ammunition and stores carriers and used for resupply of front line infantry, in which capacity they were

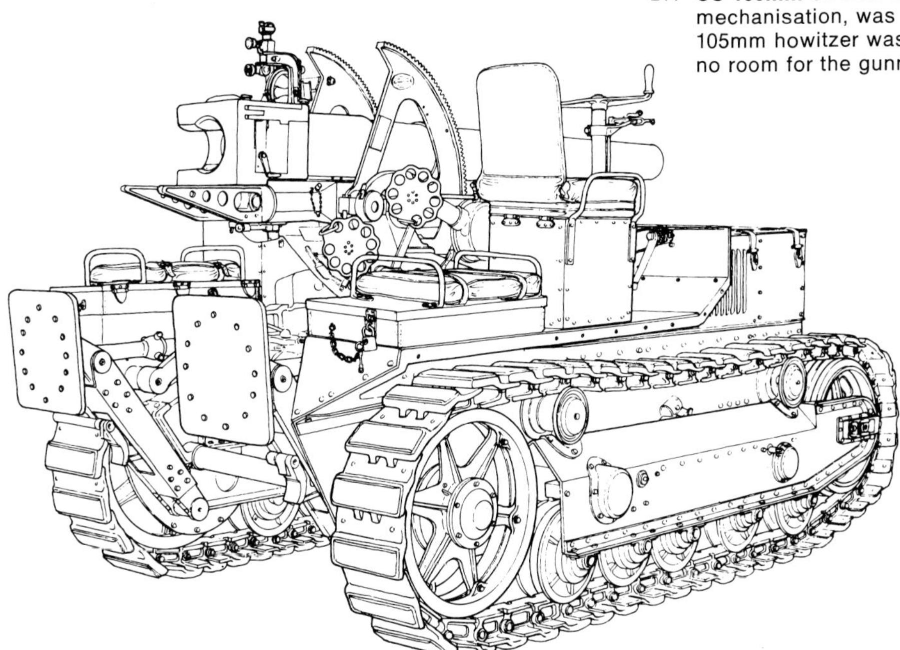

211 **US 105mm on Holt SP Mounting.** America, with her emphasis on mechanisation, was soon trying to mount guns on tracks. This 105mm howitzer was an early attempt at self-propulsion but had no room for the gunners or ammunition.

far more useful.

Although the official French line was to develop small infantry-accompanying tanks, the private manufacturers began to design proper self-propelled guns. Schneider and St Chamond, both produced a small number of 155mm guns on purpose built chassis which were used in the latter stage of the war with some success. Their designs show an interesting diversity of opinion as to motive power. Schneider elected to use a standard gasoline engine on the mounting, driving the tracks through the usual system of gears, while St Chamond divorced the engine completely from the vehicle by using a second small tracked chassis as the power supply. This carried a gasoline engine driving a generator: the tracks of both the generator carrier and the gun carriage were driven by electric motors, and the two vehicles were connected by a rigid tow-bar carrying an electric cable. Thus the driver of the generator carrier controlled the speed and direction of the assembly, the driver of the gun carrier doing nothing more than steer his vehicle in the wake of the generator.
These guns, few as they were, were seen in action by the US Army, and with their great readiness to believe in the internal combustion engine, they seized the idea. The Westerveldt Board in its 1919 report said firmly that 75mm and 155 SP guns "should be immediately developed to the utmost, paying particular attention to mobility and lightness, consistent with strength and stability". A self-propelled 8 inch howitzer, using a British piece on an American designed chassis was ready for production as the war ended, but the contract was cancelled at the Armistice, and in the immediate postwar years a large number of experimental self-propelled gun and howitzer designs trundled about American proving grounds, though

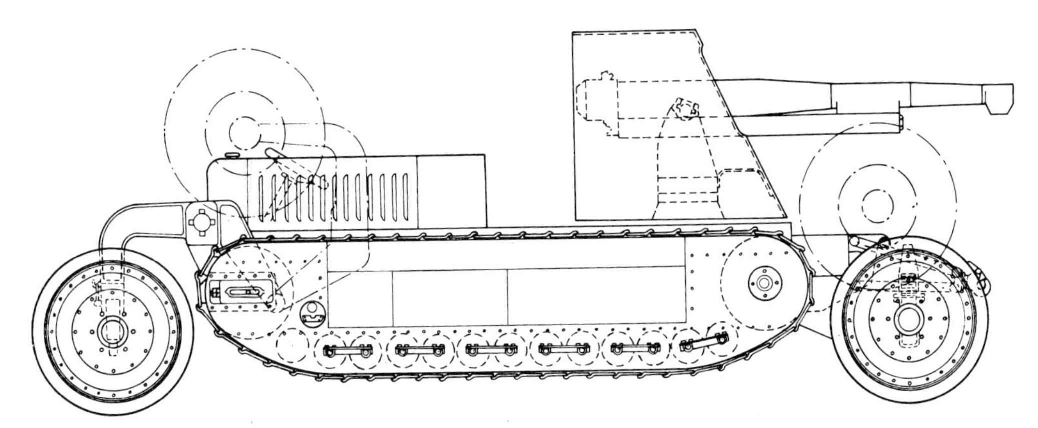

212 **Experimental St. Chamond.** This 75mm SP was developed privately during the 1920s by the St. Chamond company to the design of M. Rimailho, a well known French ordnance engineer. It could be wheeled for road use or tracked for cross-country, with the wheels retracted as shown. It was never adopted for service but gave rise to a number of imitations in later years.

none of them ever got past the gate to be accepted by the Army. The usual starting point was a Holt Caterpillar tractor chassis (211) but other designs were explored, and Walter Christy later to become famous for fast tanks and unusual suspension designs, was also active in the SP gun field; it is believed that he had a hand in the experimental 120mm AAgun design for example.

In Britain too there was an increase in gunner interest; one factor was the memory of Flanders mud which led to some interesting experiments with tracked suspension for towed guns, particularly for the short-lived 3.6 inch AA gun, and it was a short step from that to adding a meter to produce an SP. Eventually the standard Vickers tank was taken as a chassis and a special version of the 18 pounder field gun developed to fit on top. Two designs were produced for evaluation (214) one having the ability to elevate to 85 degrees and thus function as a dual purpose field and AA gun. Although officially called the "18 pounder Mark 5", these designs have always been known as the "Birch Gun" after the Master-General of the Ordnance responsible for their introduction. They became part of the Experimental Mechanised Force assembled for trails in the early 1930s but they were short-lived. Arguments developed as to whether these strange hybrids were tanks or guns, and whether they should be manned by troopers or gunners. Interest waned, money grew shorter, and eventually the whole mechanised force was disbanded and the Birch guns were scrapped.

This sort of tentative foray was repeated in other countries, and it was not until the Second World War began that the SP gun began to be considered seriously. The Germans were the first to put numbers of SP guns into action. Their experience on manoeuvres and in the Polish campaign showed the desirability of having heavy firepower accompanying the infantry assault so as to be immediately available to deal with defensive posts or guns too difficult for foot soldiers to counter. Tanks were called on to perform these chores, but the contemporary German tank gun was a small caliber high velocity weapon with insufficient shell power, consequently the German development began as a mobile heavy caliber assault gun, for delivering heavy fire at short range. Some were no more than field howitzers top carriages dropped bodily onto a suitable chassis and enclosed in armour plate, while others were slightly more purpose-built. Soon their utility was extended by fitting suitable sighting arrangements so that when not assisting in the assault they could be deployed as indirect-fire field artillery.

The Germans were also among the first to see the virtues of self-propelled anti-aircraft guns for the protection of troop columns on the move, though their air superiority during the early stages of the war lulled them into a false sense of security in this area. As a result their early efforts were not produced in volume, and the majority of those which were produced found their employment largely in Russia, where the vast distances involved stretched air protection to its

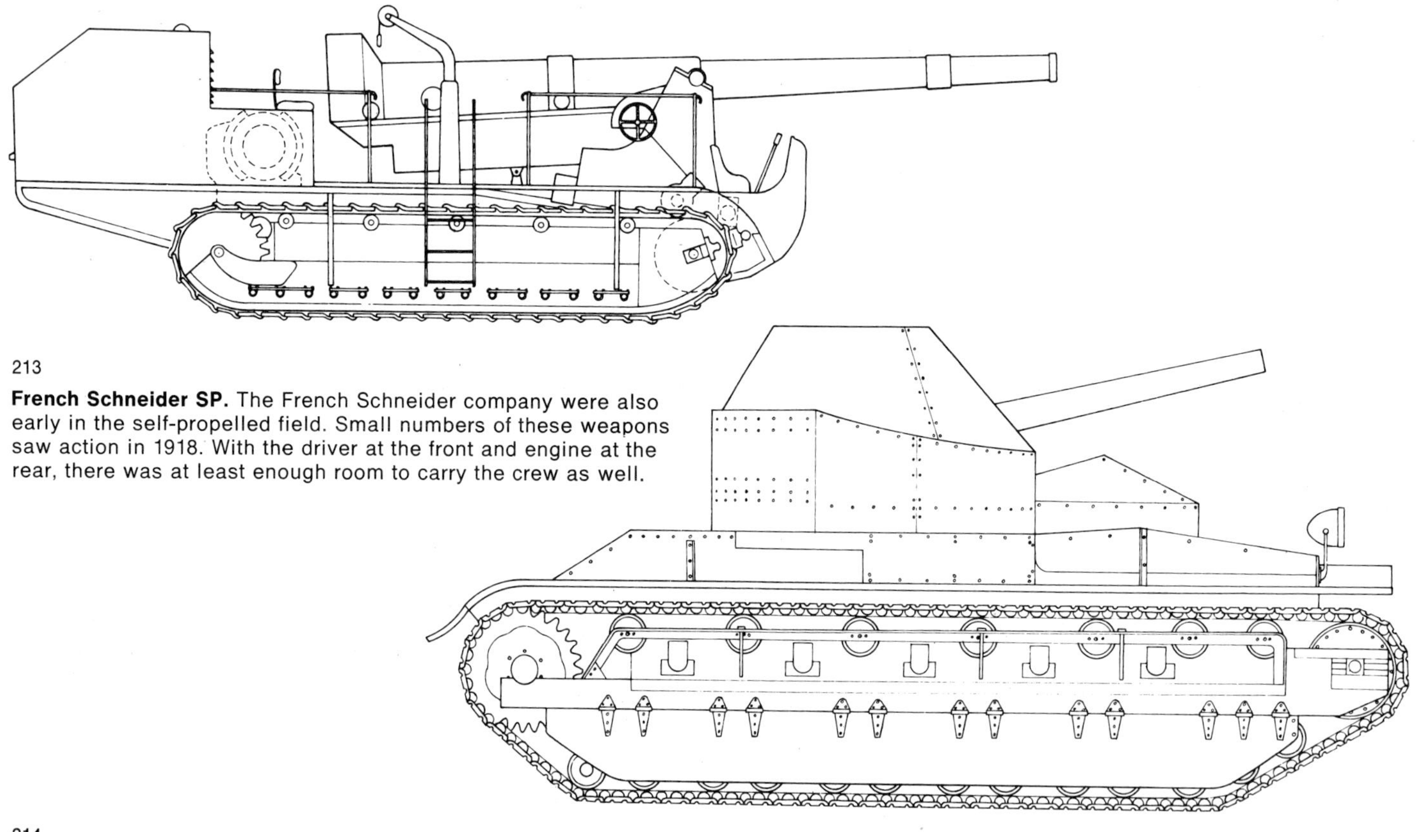

213

French Schneider SP. The French Schneider company were also early in the self-propelled field. Small numbers of these weapons saw action in 1918. With the driver at the front and engine at the rear, there was at least enough room to carry the crew as well.

214

The Birch Gun. In the late 1920s the British Army took a flyer into the mechanisation field and produced this 18-pounder SP gun for use in their experimental armoured division.

215

St. Chamond SP. Another French attempt at self-propulsion, the St. Chamond used an electric drive with current supplied from another vehicle by an umbilical cable. Note the use of rear jacks to stabilise the suspension during firing. Photo shows three variations of the St Chamond with the ammunition carrier on the extreme left.

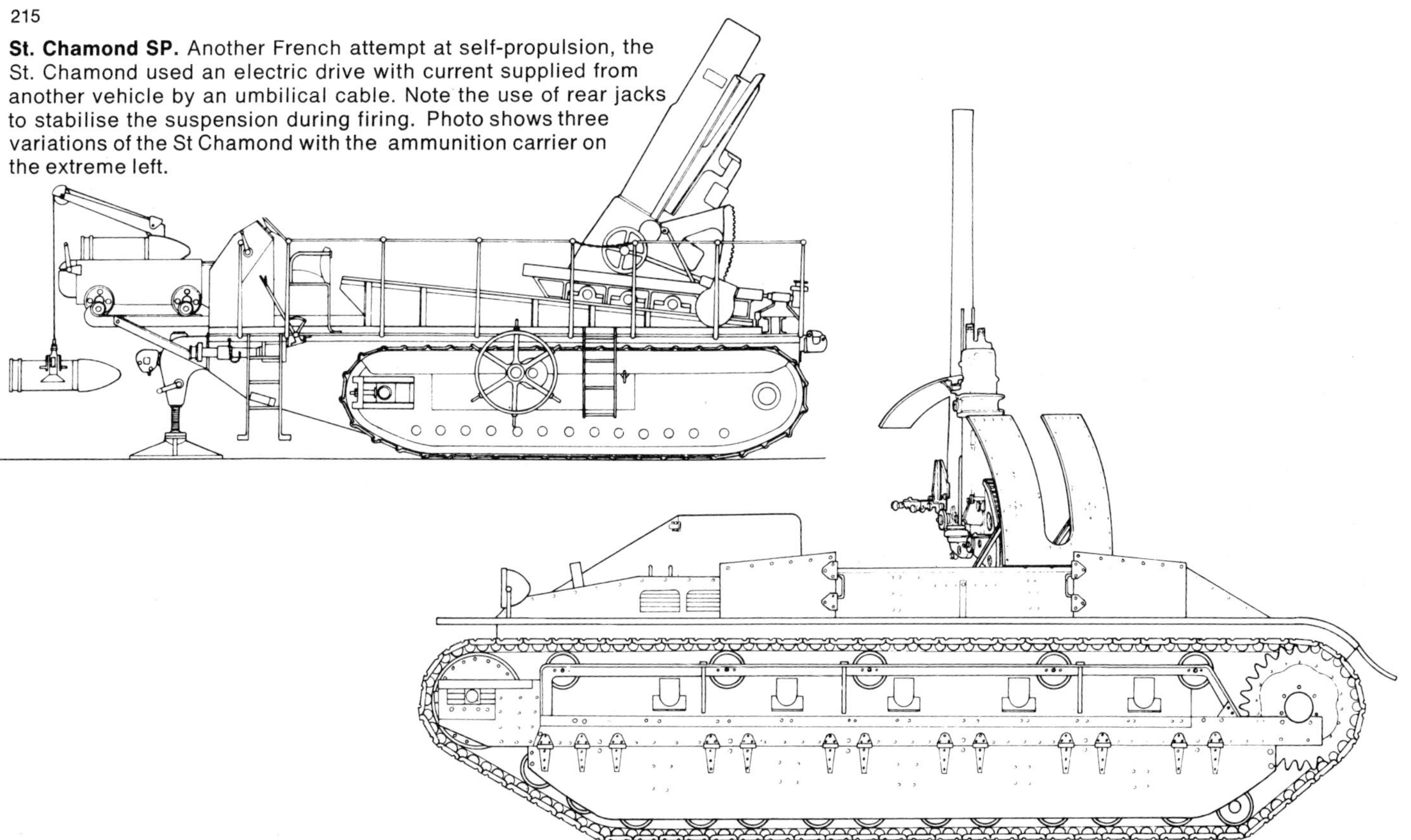

216

High Angle Birch Gun. Another variation on the 18-pounder SP design, this had the gun mounted so that it could function either as a field piece or as an anti-aircraft weapon.

limit. The most common equipment was the light 20mm or 37mm Flak guns mounted on the back of a three-quarter tracked artillery tractor, though some few were placed onto tank chassis. Later versions used the twin and quadruple gun mounts and two which achieved the dignity of a name were "Möbelwagen" (furniture van) and "Kugelblitz" (Ball lightning). Parallel development on the Allied side was confined to the American adoption of multiple .50 inch heavy machine guns, sometimes combined with a 37mm gun, on half track mountings. Britain toyed with mounting the 20mm Polsten cannon on vehicles and trailers, but these were relatively few and the principal British innovation was the mounting of the 40mm Bofors on a wheeled chassis. All these equipments had the ability to produce a surprising volume of fire at short range, and in addition to their anti-aircraft role, they were often in demand as extempore assault guns to help the infantry in small actions.

With the increasing use of armoured divisions it became necessary to provide cross country mobility to the supported tanks, and this led to increasing work on SP gun designs. Another special incentive, so far as Britain was concerned, was the likelihood of invasion in 1940 and 1941, in which mobile weapons seemed to be desirable properties for dashing rapidly to threatened points. It was this aspect which was dealt with first in Britain, and numerous odd designs of armoured gun carrier were rapidly thrown up. From the lightweight "Beaverette", a car chassis covered in slabsided armour and mounting a light machine gun (named after Lord Beaverbrook who was largely instrumental in getting them into production) to such things as "Deacon", a 6 pounder on the back of a Thornycroft truck (220) and "Firefly" a 6 pounder in an armoured-car-like configuration (219) most of these devices were wheeled, since this gave a better chance of rapid production than a tracked design. Most of these inventions never got past the prototype stage, though "Beaverettes" were retained for airfield protection throughout the war, and a handful of "Deacons" appeared briefly in the Middle East in 1941 and 1942.

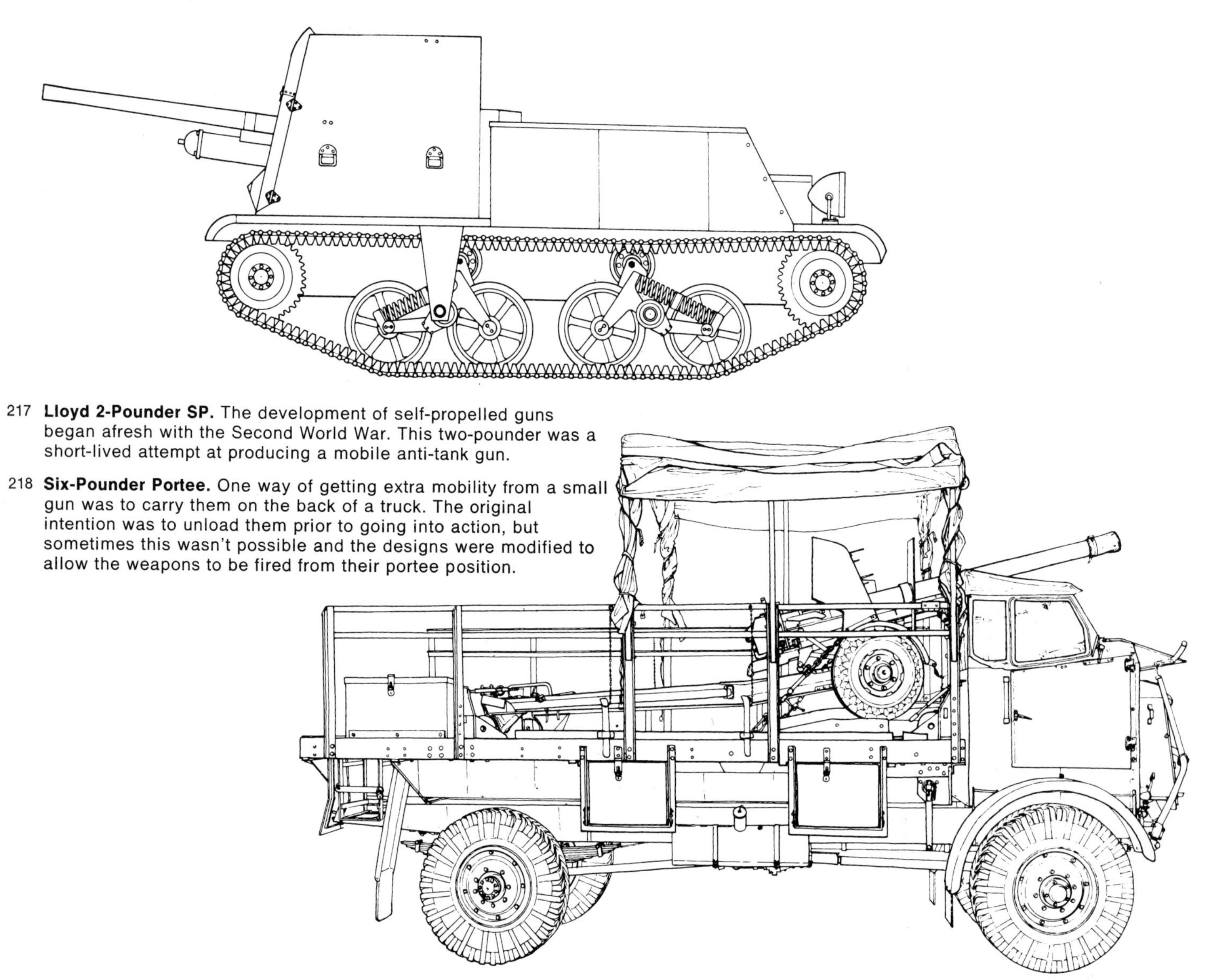

217 **Lloyd 2-Pounder SP.** The development of self-propelled guns began afresh with the Second World War. This two-pounder was a short-lived attempt at producing a mobile anti-tank gun.

218 **Six-Pounder Portee.** One way of getting extra mobility from a small gun was to carry them on the back of a truck. The original intention was to unload them prior to going into action, but sometimes this wasn't possible and the designs were modified to allow the weapons to be fired from their portee position.

During the Desert War the problem facing anti-tank gunners was the questionable reliability of their guns when confronted with action after having been towed over miles of indifferent roads or across the desert itself. The vibration and hammering taken by a gun carriage under these conditions is much greater than the casual onlooker might imagine; indeed there is one much-vaunted gun in service today which is almost guaranteed to start shedding pieces after thirty miles on a smooth road, let alone trying to take it cross-country. So the Desert gunners developed the idea of heaving the gun onto the back of the towing vehicle and carrying it piggy-back up the line until battle was joined. It could then be dropped off, hooked up and towed in the usual way. Inevitably, of course, Finnegan's Law working full blast, somebody miscalculated and arrived in the midst of the battle with his gun anchored on to the top of the truck, and that was how the practise of firing it from the truck was invented. Eventually specialist trucks were designed and built to "portee" anti-tank guns (218), some of them having the designed ability to fire from the portee position. But once the guns began to increase in size from the two-pounder and six-pounder class, the portee system had to be abandoned.

When it came to tracked designs, the production difficulties were so great that the only way to get a design even considered was to start with a chassis already in production: and the only one in sufficient quantity to allow experiments was the infantry carrier. This armoured utility vehicle came in a number of variations—the Universal, The Carden-Lloyd and so forth—but they were all very similar, and designers were soon at work shoe-horning a variety of ordnance into them. The 2 pounder anti-tank gun was mounted in a number of configurations, from alongside the driver to being placed in a turret on the rear deck (217), and when the 6 pounder gun came into production various methods of attaching were tried. Eventually the 25 pounder was even installed though where the detachment and ammunition were to go appears not to have bothered the designer overmuch. Apart from a few 2 pounders none of

219 **Firefly.** The British Morris Firefly Tank Hunter was an attempt to bring the six-pounder to the tank in cases where the tank was reluctant to come to the six-pounder. It never entered service.

220 **Deacon.** Another attempt to hunt tanks down was this six-pounder wheeled SP called "Deacon".

221 **Lloyd 25-Pounder.** One idea which didn't work was this British attempt at mounting the 25-pounder into an infantry carrier. No traverse, no ammunition storage, and the poor gunner, working like the proverbial one-armed paper-hanger, had to load, lay and fire it all by himself.

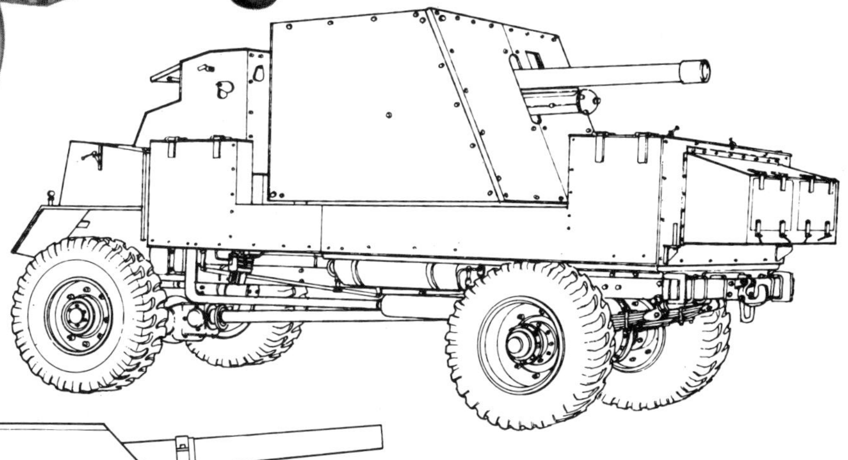

222 **Priest.** The US 105mm Howitzer Motor Carriage M7. The ultimate solution to the SP problem is, of course, a full-tracked chassis, and this design which used the M4 tank chassis as a basis proved to be highly successful.

223

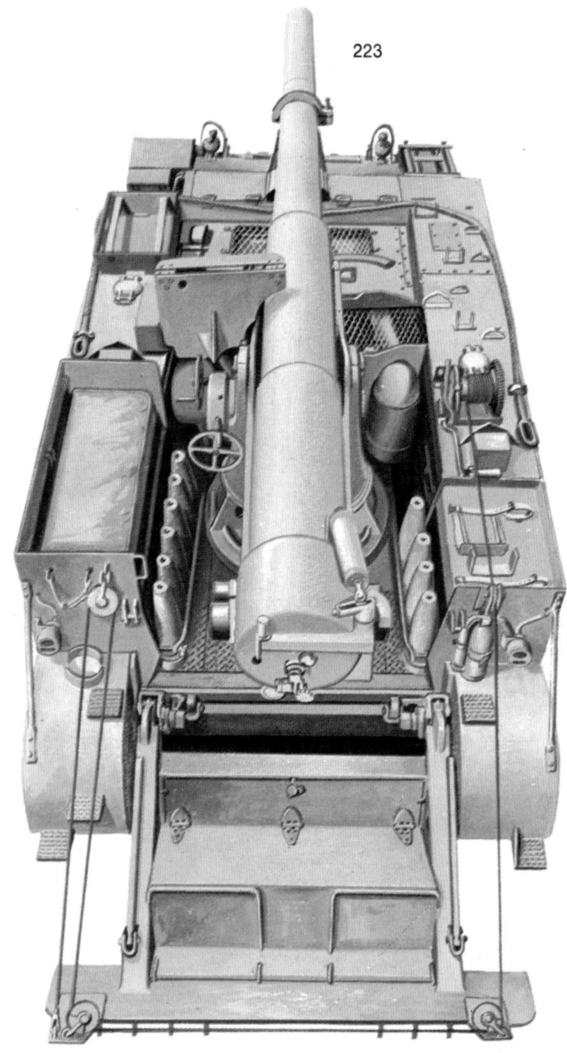

223 **US 155mm Gun Motor Carriage M12.** One way of using up old guns is to turn them into SPs. This is how the US Army got extra service out of the M1918 gun when it had been replaced as a towed weapon by the improved M1.

224 **Sexton.** The Canadian "Ram" tank was not a success, but when converted and with the 25-pounder gun installed it became one of the best SP field guns of the war years. This version is fitted with skirting plates over the tracks, an uncommon accessory.

225 **Archer.** This is the British 17-pounder mounted into a Valentine tank chassis. It had to fire over the rear, which made manoeuvre difficult, but it turned out to be a highly effective weapon.

226 **US 75mm Howitzer on Half-Track.** In the early days of World War Two the US Army tried mounting guns on to half-track scoutcars in order to give them sufficient mobility to keep up with armoured and cavalry formations. This was the 75mm Howitzer version.

227 **US 105mm on Half-Track.** The 105mm Howitzer was also half-track mounted. Although not standardised, numbers were built and used in North Africa in 1942.

223

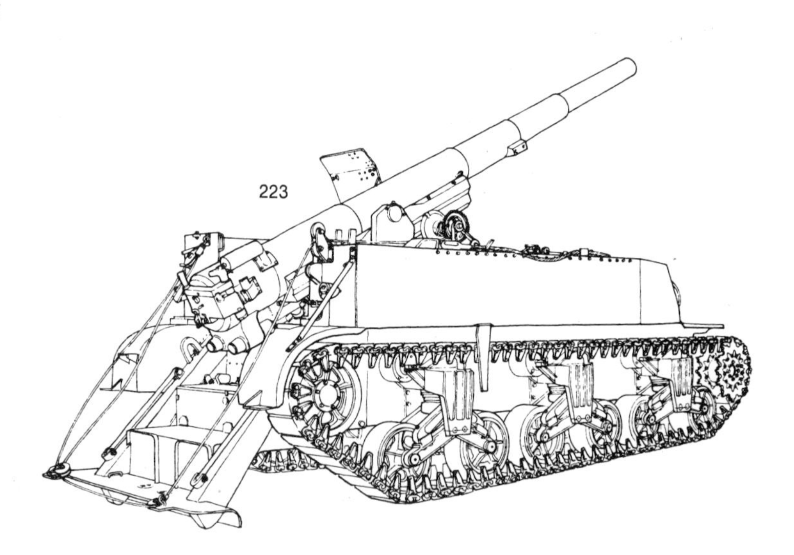

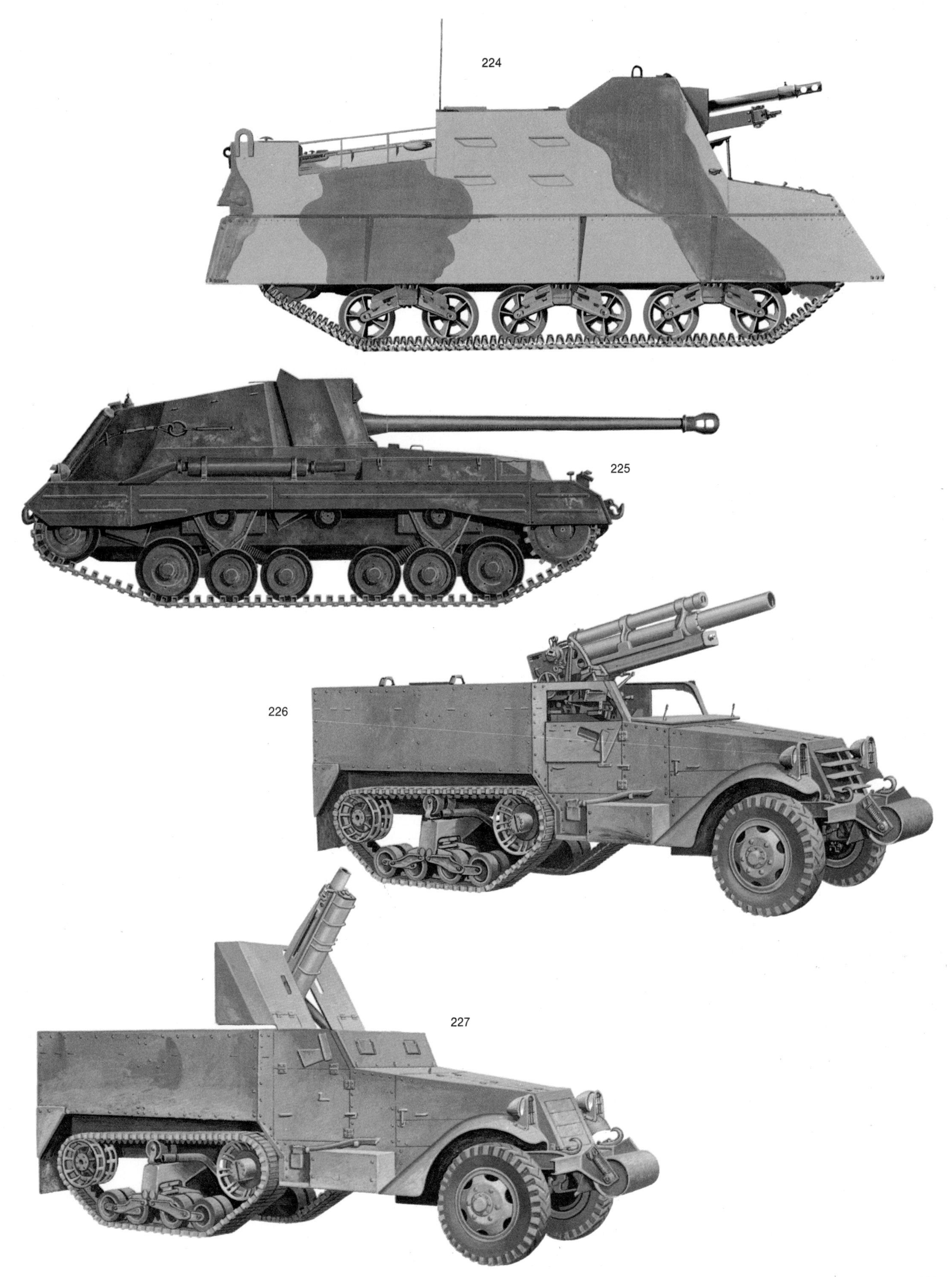
224
225
226
227

these carrier based designs ever appeared in quantity. The Americans were having their silly season at about this time, but they were concentrating on the problem of keeping up with the armoured divisions. Not having the carrier to play with, they took the equivalent, the halftrack scout car and tried grafting the 75mm gun, 75mm howitzer and 105mm howitzer on to it. As well as their employment as field weapons these were also used as SP anti-tank guns in the tank destroyer units, since the contemporary American anti-tank tactic was based on an offensive hunt-and-shoot doctrine. Numbers of these halftrack SP's saw service in the North African campaign in 1942, but they were avowedly a stopgap until tracked vehicles were designed and built.

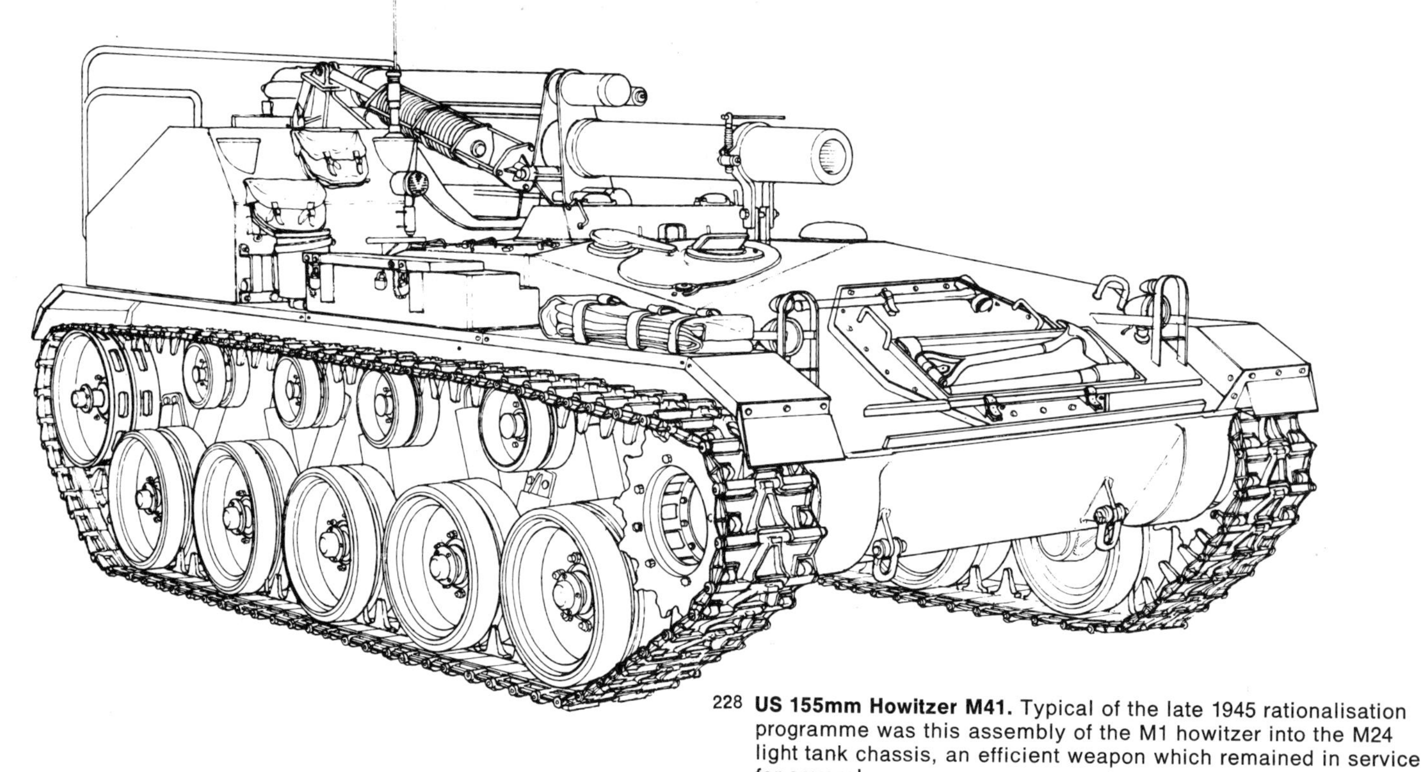

228 **US 155mm Howitzer M41.** Typical of the late 1945 rationalisation programme was this assembly of the M1 howitzer into the M24 light tank chassis, an efficient weapon which remained in service for several years.

229 A fully automatic Swedish 57mm AA self-propelled gun in the firing position and, opposite, ready for a rapid move. This experimental turntable mounted gun shows some unusual solution to a variety of mechanical problems. Photo: Bofors A.G.

230 **Churchill Special.** A Home Defence weapon developed in Britain during the Second World War was this adaptation of old 3 inch AA gun barrels into the Churchill tank. Only fifty were built.

231
This Sexton is carrying a non clerical trophy. Was he hit or run over?

Another American idea was to use the SP gun as a method of finding a use for obsolete weapons which had been replaced by more modern designs of carriage but whose ordnance was still capable of useful service. The 155mm gun M1917 had been superseded by the M1 in 1938, a much improved weapon, but the hundreds of M1917 barrels which remained were perfectly serviceable and ample stocks of ammunition were available. By mounting the gun onto a converted tank chassis the M12 gun (223) was produced and numbers of these saw service during the war. In order to provide the stability so desired by the Westerveldt Board, the chassis was fitted with a bulldozer blade at the rear which could be dropped to the ground and the vehicle reversed against it to give a solid recoil platform and relieve the suspension of some of the stresses. This arrangement has been a feature of almost every US heavy SP design ever since.

The 105mm halftrack was replaced in the armoured divisions by a design using the chassis of the M3 tank as its starting point. As a tank the M3 was unsatisfactory due to its high silhouette and sponson-mounted gun gave a restricted field of fire, and when better designs became available the M3 was re-worked into an open-top design into which the top carriage of the standard towed howitzer was fitted. Though the resulting weapon, the Howitzer Motor Carriage M7, was a serviceable and well liked piece of equipment, it suffered from its mixed origins, being overweight and with the guns elevation and traverse restricted so that the full performance of the weapon was not utilised. Issued in April 1942 the first off production went to the British Eighth Army in the Western Desert where, due to the pulpit like machine gun mount, they were promptly christened "Priest" and started the British tradition of giving ecclesiastical nicknames to SP guns.

On the other side of Africa the British 1st Army, later in 1942, found themselves using the first British SP field gun the Valentine-based 25 pounder nicknamed "Bishop"., This like the M7 was the marriage of an of an outdated tank chassis to the standard field gun, but it was a good deal less successful than the M7. Whether the designer had been influenced by early Russian designs is hard to say but the Valentine Chassis was surmounted by a large square box in which the gun was mounted. This mounting so limited the guns elevation that the maximum range was reduced by nearly 4000 yards. After the North African campaign had finished, an official report noted that "nothing good can be said of the Valentine 25 pounder". But the M7 had pointed the way; instead of trying to ape the tank, using a turret to gain overhead protection, the SP gun was to become an open-topped vehicle in which protection was sacrificed to give mobility to the chassis and flexibility to the gun. After all, the SP was simply a matter of providing a field piece with mobility, not developing an armoured fighting vehicle. While the Valentine held the ring and gave the gunners a chance to develop tactical doctrines and drills for the new weapon, its replacement was on the way.

The Canadians in a pardonable excess of national enthusiasm had developed their own medium tank, the Ram. This was basically the US M4 Sherman with a few local variations, and as a battle tank it was an under gunned non-starter which was obsolescent before it got into production. It had however a reliable and well built chassis and this was now re-designed to provide an open-top hull into which the 25 pounder gun could be fitted (224). This mounting gave ample traverse and full elevation so that all the performance of the gun was available, and under the name of "Sexton" it became a reliable and well liked weapon, remaining in service with the British until scarcity of spare parts retired it in the late 1950's.

The Valentine-based "Bishop" was forthwith retired, but

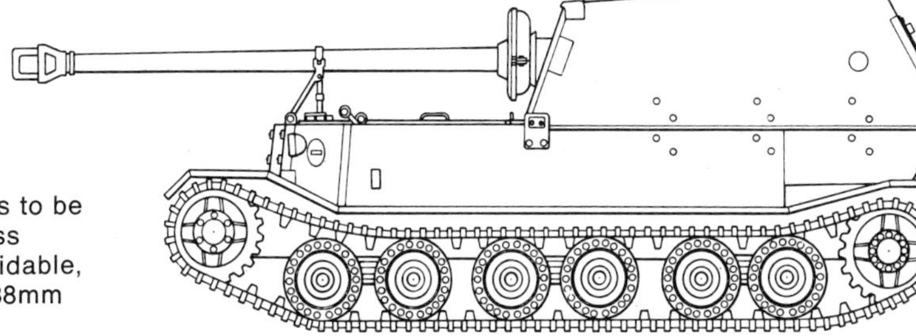

232

Ferdinand The Elephant. Designed by Dr. Porsche, this was to be the tank hunter to end all tank hunters, but it turned out less successful than had been hoped. On the move it was formidable, but once stopped it was a sitting duck, since its powerful 88mm gun had practically no traverse.

233 Two American 3 inch M10 SP guns firing at night during the Italian campaign of 1944.

the Valentine chassis found another role when it was used as the basis for a 17 pounder SP anti-tank gun. Due to various structural requirements the gun had to be mounted firing to the rear, with limited traverse, but the weapon turned out to be nimble and manoeuvrable, and numbers were used in Europe in 1944-45; it remained in service for some time after the war. This weapon was christened "Archer" which is at variance with the churchmen which preceeded it and is explained by the British practice of refering to armoured fighting vehicles as "A" vehicles, and non-combat vehicles as "B" vehicles. "A" vehicles thereafter acquired names beginning with "A"—Avenger and Alecto for example.

In America the shortcomings of the M7 were realised and early in 1945 it began to be replaced by a new design, the M37. This was similar in general outline but was based on the chassis of the M24 tank, a product of a rationalisation programme in which selected chassis designs were to form the basis of both tanks and SP guns to simplify supply and maintenance. The gun and its mounts were also redesigned to use a concentric recoil system, in which the gun barrel passes through the recoil cylinder and actually acts as its own piston. This gives a much cleaner and compact design, (it had first been developed for use with tank guns to cut down the clutter in the turret) and, this with other smaller modifications, allowed the M37 to

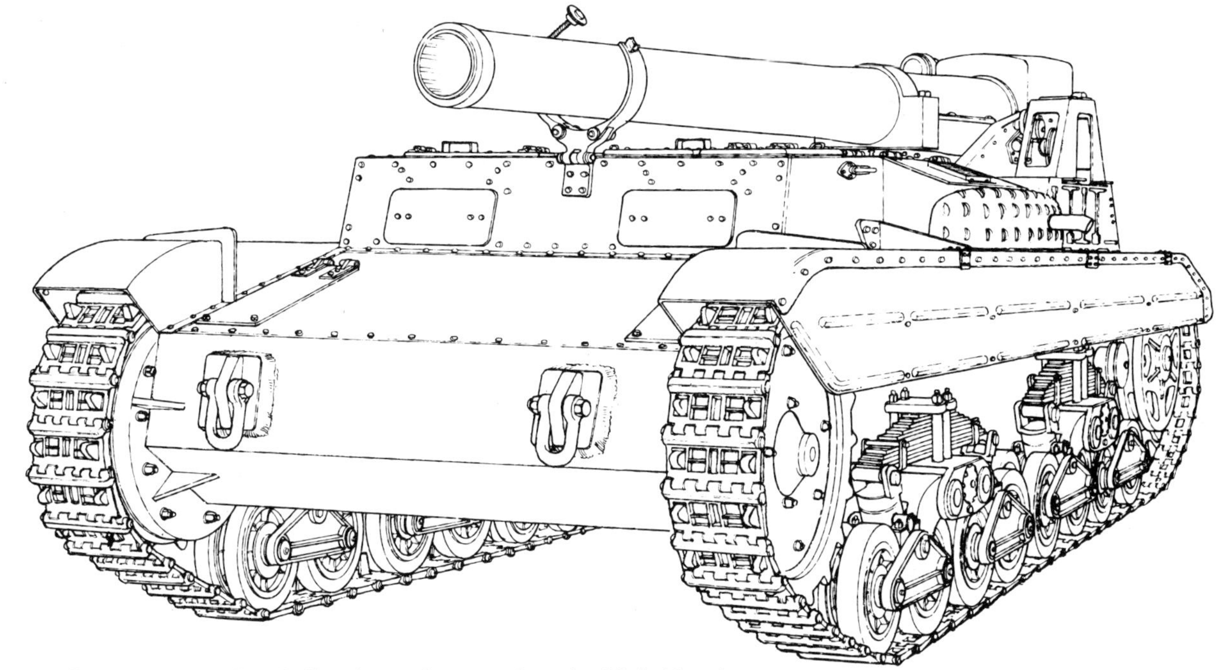

234 **Semovente 149.** The Italian Army also went into the SP field and one of their better designs was this 149mm. With no protection for the crew and precious little carrying space for either them or the ammunition, it nevertheless had a good mobility.

235 **17-Pounder M10.** The American M10 Tank Destroyer was originally fitted with the 3 inch gun. It was later up-gunned by the US Army substituting a 90mm and the British a 17-pounder.

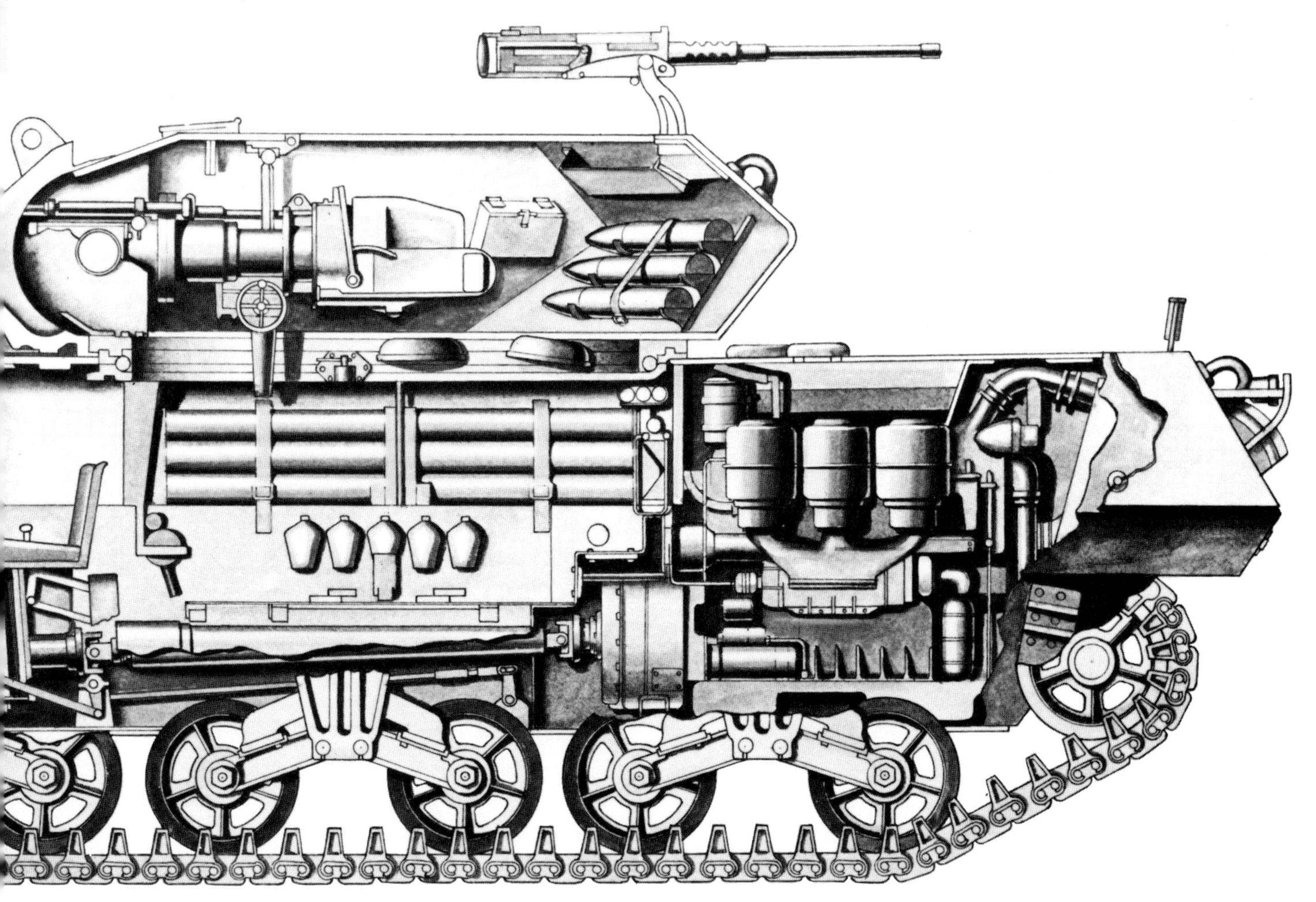

236 **Grizzly Bear.** A German assault gun of impenetrable aspect, the Grizzly Bear provided close-in support for infantry attacks.

237 **German Sturmgeschutz.** This cutaway drawing of the 75mm StuG III well illustrates the complexity necessary to get a simple gun close to its target on the modern battlefield.

238 **JSU 122.** The Soviet tactic of frontal attack placed great dependence on the assault gun and this model was the result of building the 122mm divisional gun into the chassis of the Josef Stalin tank.

239 **JSU 152.** An even heavier weapon, this Josef Stalin conversion carried the 152mm howitzer. As with the 122mm model, it could be used as an assault gun to shoot the infantry into their objective or it could be deployed in the usual artillery fashion to provide indirect supporting fire.

elevate its gun to 45 degrees and thus use the full performance. Few of these actually reached the front-line troops before the war ended, but it was to serve the US army for many years afterwards.

The German army developed innumerable SP's. When a tank was outclassed its chassis was invariably called on to do duty as a gun mounting, though the assault gun concept still dominated German thinking. Captured vehicles too were pressed into service—the French Lorraine and Somua and the Czech 1938 being popular in this role. The imposing three-quarter-tracked Krauss-Maffei artillery tractor also found employment as a platform for light anti-aircraft and anti-tank guns for column protection particularly on the Eastern front One of the less happy (232) German developments was the "Ferdinand", a massively armoured vehicle carrying a high velocity 88mm gun. This 68 tonner looked formidable but the gun had limited traverse and elevation—it was a tank destroyer pure and simple. It stemmed from a Porsch design contestant in the Tiger tank development programme. Turned down as a tank, Porsch redesigned it. Ninety were built, and sent to war on the Eastern front. Few survived their first engagement. As long as the enemy was in front, "Ferdinand" did everything it promised and was invincible, but once an attacker got to the side of the limited field of fire they were sitting ducks. Those which survived were re-christened "Elephant" by the troops, a sufficient comment on their ability.

So far as field pieces were concerned, almost anything went. The first attempts were simple and stark; the Imperial War Museum in London have an interesting specimen of unknown parentage which consists of the chassis of a Panzerkampfwagen III onto which a slab-sided superstructure has been grafted and a complete 105mm field howitzer, minus its wheels, dropped in and bolted down. In order to reduce stress on the mountings of these extempore weapons it was found preferable to use large calibre low-velocity weapons giving the desired results by shell-power, rather than choose a smaller high-velocity gun which would have demanded stronger and more complicated mounting arrangements. Thus the preferred German weapon was the 15cm howitzer, and in an attempt to reduce the mounting stresses even further, many of these were fitted with non-standard low-efficiency muzzle brakes. Another interesting point about these German self-propelled equipments, which arises from the reduced firing stresses, is that contrary to American practice in similar calibres, none of them were ever fitted with any stabilising spades or jacks, the entire firing stress being absorbed by the vehicle suspension, making it perfectly feasible to fire them on the move if necessary. It also meant that their emplacement and displacement involved nothing more than telling the driver to either stop or start up.

Mechanical ingenuity is always liable to crop up in German designs, and an odd variation on the SP gun was the "Heuschreke" or "Grasshopper". This was a tracked chassis with a turret, resembling a tank which carried on its rear deck a pair of folded crane arms (their resemblance to a grasshopper's legs gave rise to the nickname). The idea was that this vehicle would trundle across bullet-swept ground to a selected fire position and then, with the crane arms, lift off the turret and emplace it on the ground as an armoured pill-box. The carrier could then trundle back to repeat the performance somewhere else with another turret or merely park in a concealed position until required to remove the pillbox again. Krupp did most of the development on the equipment but it was still in its trial stage when the war ended.

Another German development was the Waffentrager (Weapon Carrier) which harked back to the British 60 pounder Gun Carrier of 1916. A tracked chassis carried an armoured body, open at the rear so that a field or anti-tank gun could be run up ramps and clamped in place. Once aboard it could be fired from the vehicle and used as a self propelled gun, or it could simply be transported across country and dropped off at any selected location to fire from its wheels in the usual way. Again, the development had not reached the production stage when the war ended.

The Russians have always been artillery conscious, and they embraced the SP idea whole heartedly. As with the Germans, their first approach was the assault gun, largely due to their limited tactical doctrine—to go for the enemy before he went for them. So their Josef Stalin-based weapon, mounting 122mm and 152mm howitzer were essentially direct fire brute force weapons to be used as bludgeons whenever Ivan ran up against something too hard. But later, when set piece battles developed on the route to Poland and Germany these weapons took up indirect fire positions to provide supporting fire in the accepted manner. Then, as the defenses began to give way, they could lower their muzzles to point blank and drive off to shoot the infantry at close range. The Soviet equivalent of the Sexton and M7 was the SU 76, the standard 76mm field gun mounted onto a specially built chassis. While a relatively useful weapon it was never as popular or as widely used as the SU 122 and 152 being deficient in shellpower for the assault gun role, and soon outmatched by increases in German armour.

An early Russian attempt at self-propulsion was the Klim Voroshilov II. The Klim Voroshilov was a heavy tank, named in honour of the then Defence Minister, and it was armed with the same 76mm gun as the T34 tank but about twice as heavy due to having much thicker armour. Battle experience showed that the T34 was the better all-round weapon, and so the majority of KVs were re-worked into KVIIs by fitting a new turret carrying a 152mm howitzer. In order to allow the crew sufficient room to serve this weapon, the turret had to be a large box-like structure which put the weight up to 57 tons and produced a very vulnerable silhouette. Once the lower and more workman-like Josef Stalin based designs were developed, the Klim Voroshilov was retired.

The U.S. Army had by now replaced the M12 SP 155mm gun with the M40, which was the M1 gun and the top carriage mounted onto a modified M4 Tank chassis. This enabled the full range of the M1 gun to be utilised instead of the SP guns being restricted to the limited performance of the older weapon. The 155mm Howitzer M1 had also been installed on a M24 chassis to produce the M41 SP

Howitzer, (228) and this weapon moved a further step away from the early all-armour concept as very little more than a chassis with a howitzer on top, totally devoid of armour protection. As the war entered its last stages the US Army was concerned with the defence likely to confront them in Okinawa, Iwo Jima and on the mainland of Japan. The Japanese soldier had shown himself to be an enterprising builder of tough defensive works ingeniously constructed and invariably covered by interlocking areas of fire so as to make direct assault a hazardous business. The 155mm SPs, useful weapons as they were, were felt to have insufficient punch for long range bunker busting and designs were begun to place the 240mm Howitzer M1 on a chassis built up from M26 components. Christened "King Kong" by some unknown wit, this turned out successful and was then partnered by a similarly mounted 8 inch gun. A small number of these formidable weapons were built and were en route to the Pacific when the Atomic bomb rendered them superfluous overnight.

As befitted a race who produced some of the world's best engineered cars, the Italians came up with several designs of SP guns, but somehow the vital spark which produced the Alfa Romeo and the Lancia Dilambda was missing, and the majority of the designs were unmemorable. The only one to be used in large numbers was the 149mm Semovente (234) used in the desert war and also adopted by the Germans during the Italian campaign.

After the war ended there was the usual breathing space while projects were assessed and many of them abandoned, and then the development of a new generation of SP guns began. But now the wheel had turned full circle and the accent was again on protection—not against gunfire and shell splinters this time, but against radiation, flash nuclear fallout and bacteriological and chemical agents. Now the ideal seemed to be a sealed and pressurised container in which the gun detachment would be safely preserved against all these insidious threats and still capable of operating their guns. US designs began to take on the appearance of displaced ship's turrets on tracks (251) and the mechanical complexities of power elevation and traverse power ramming, ammunition hoists and even the location of the driver in the turret, led to

240

King Kong. Developed as a potential bunker-buster for use in the invasion of Japan, 115 of these 240mm Howitzers T92 were scheduled to be built, but the end of the war caused the programme to be cancelled and only a handful were made.

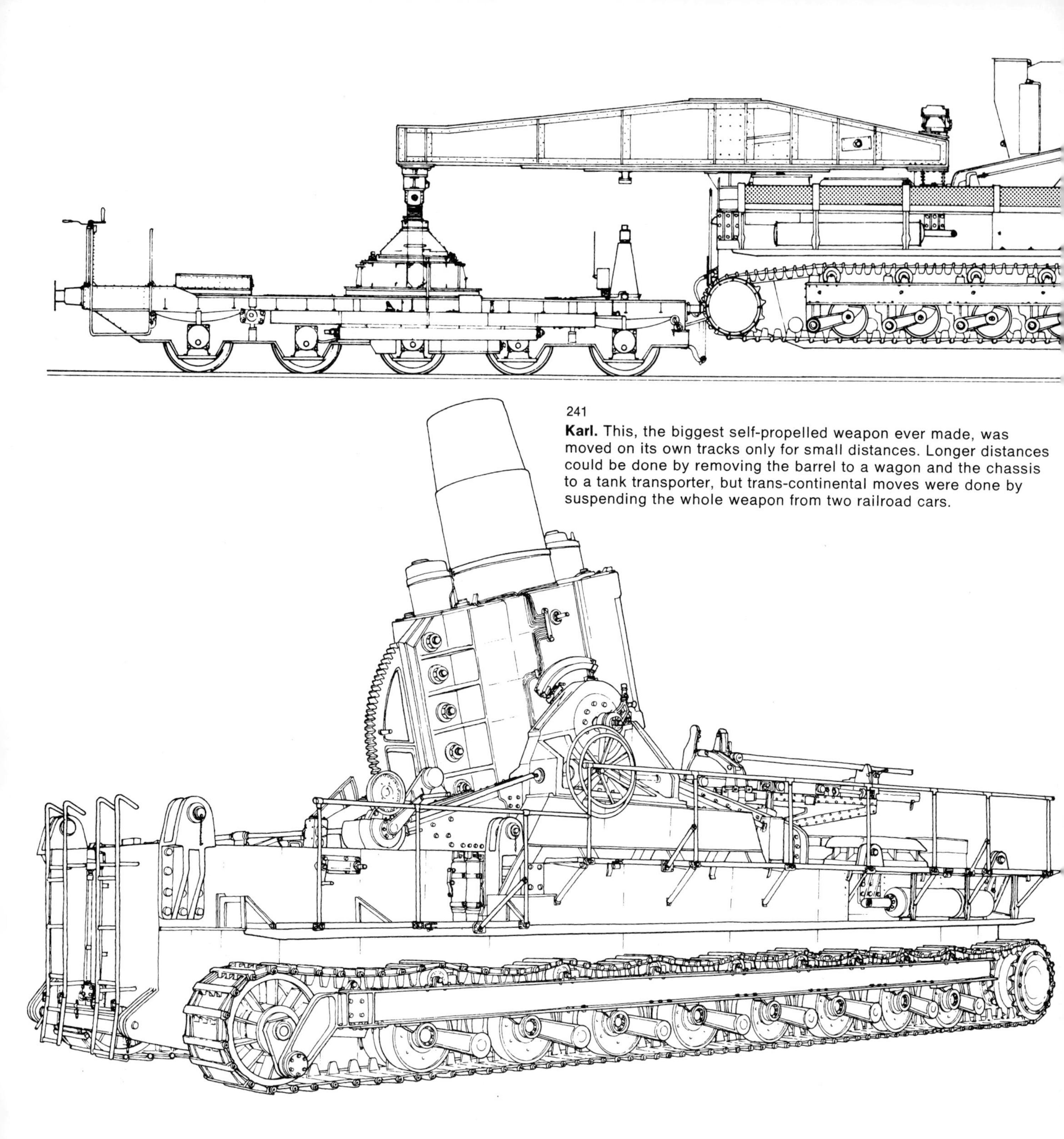

241
Karl. This, the biggest self-propelled weapon ever made, was moved on its own tracks only for small distances. Longer distances could be done by removing the barrel to a wagon and the chassis to a tank transporter, but trans-continental moves were done by suspending the whole weapon from two railroad cars.

weapons which needed mechanics rather than gunners to keep them in operating order.

The British were satisfied with their 25 pounder Sexton for several years and did little to develop SPs until the middle of the 1950s. Then, taking their well-tried Centurion tank as a basis they replaced the standard 20 pounder gun with the 25 pounder in a solid structure replacing the turret to produce a fully protected SP field gun. With the Centurion turning the scale at 50 tons, the general effect was described by one observer as comparable to putting a man with an air rifle on a double-deck bus and the project was abandoned in favour of mounting a 5.5 inch medium gun in the same way. This made more sense but before it could be perfected a NATO agreement on standardisation of alibres agreed on 155mm as the common General Support weapon, and the British came to the sensible conclusion that it would be cheaper and easier to purchase American weapons than to persist with their own design.

By this time the Americans seemed to have become disillusioned with their complex turreted SPs and had reverted to simplicity. The 155 How M41 had been replaced with the M44, the same ordnance in a newer chassis, with power

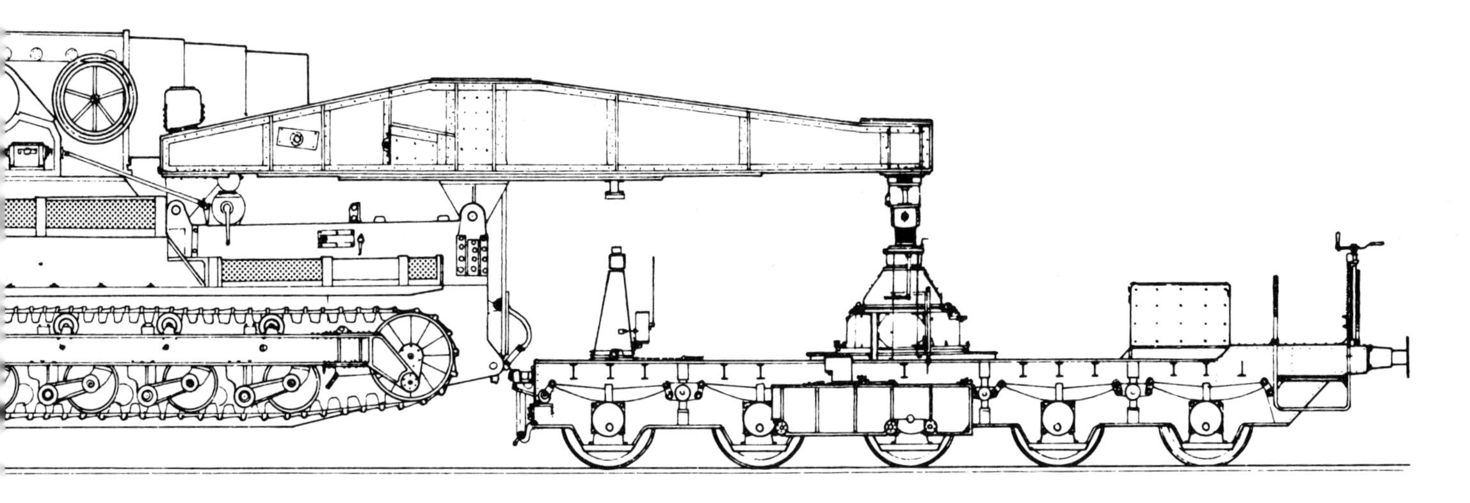

controls and the driver in a commanding position alongside the gun. Open-topped, it had side plating and a fabric collapsible cover could be erected to provide sufficient protection against nuclear flash and fallout. This weapon was adopted by Britain as their large calibre SP gun though it never achieved the dignity of a nickname. Disillusionment with the combined electric and percussion firing lock led the British Army to modify the breach mechanisms to accept their standard percussion lock, after which the weapon was given a British nomenlature as the 155mm Howitzer L8.

At the same time as the 155mm caliber had been agreed, the NATO nations had also agreed a 105mm as the caliber for their close support guns ("close support" being the "in" expression, replacing "field guns" which was "out"). A promising design of 88mm gun developed by Garrington's was well on the way to acceptance at this time: it was soon seen that up-calibering the design to 105 was out of the question and it was abandoned in favour of an SP 105mm. By 1962 this had been built and was undergoing trial and it entered service as the 105mm "Abbot" (254). Abbot, with its multifuel engine and multi charge gun is one of the best of its class in existence. It can function as a

242
A rapid firing anti-aircraft design is this German Leopard Chassis mounting twin Oerlikon 35mm guns together with radar and fire control instruments. *Photo: Krauss-Maffic.*

243

Anti-Aircraft Ram. Another use for the Canadian "Ram" tank chassis was as a basis for this self-propelled 3.7 inch anti-aircraft gun, an equipment of doubtful value. Only prototypes were built.

high velocity gun, firing HESH shell in the anti-tank role with devastating results, or it can elevate to 80 degrees to function as a howitzer. Firing HE, HESH Smoke and illuminating shells its 8 charges can provide continuous coverage to maximum range in any terrain configuration.

Continuing on their path of simplification, American SPs became more and more stark. The 90mm Scorpion anti-tank gun was no more than a gun atop a tracked chassis, with absolutely no protection for the crew but a commendibly low silhouette. The 175mm M107 (253) was similarly open, though provided with a high degree of power assistance to enable the gun's detachment to be kept to a minimum. This caliber was selected as being the optimum comparable with mobility and light weight on the one hand and the capability to deliver a useful nuclear shell on the other.

At the moment, the design of SPs seems to be undergoing a quiet spell, todays great interest being lightweight and portable towed guns, but doubtless a fresh generation will be along shortly. Also undoubtedly the USA with its unrivalled expertise in this field will show the lead, but it would be difficult to forecast what direction—protection or mobility—this new generation will go.

A sort of halfway house between the towed gun and the SP gun is the wheeled equipment having some sort of power pack tied on in order to provide "auxiliary propulsion" to use the official term. Probably the first in this class was the Straussler converted 17 pounder anti-tank gun developed in Britain in 1943. (255). Nicholas Straussler was a British engineer responsible for a number of inventions in the vehicular field, and this modification of the 17 pounder gun was intended to give the weapon mobility independently of its towing vehicle for short deplacements. The axle was extended to allow installation of an engine, transmission and drivers controls. The trail ends were mounted on a two-wheeled limber and the whole affair could then be driven, albeit slowly, from position to position. But the complexity and degree of modification needed appeared to be more than the limber mobility was worth, and the most objectionable feature in the soldier's eyes was the enormous pit which had to be dug to accommodate the enlarged weapon. As a result the Straussler design was never adopted for service.

Some years later, in the 1960s, the Soviets unveiled a 76mm gun with what appeared to be a motorcycle engine pack on the trail, driving and steering through a single dolly wheel under the trail ends (256) and later followed this with a similar design of 57mm gun. And as these words are being written, the first pictures of the Anglo-German-Italian 155mm General Support gun project have been released, showing that a similar arrangement closely related to Straussler's idea, has been adopted, with a power and control pack ahead of the axle and the trail ends supported on a two wheeled dolly. Probably the principal utility of these aux prop devices is for airborne use, where the gun can be parachute and helicopter delivered without its prime mover and the power pack will provide all the mobility necessary for operation in the restricted area of the warhead before success and expansion permits the vehicles to be delivered. Other than this,

there seems little to be gained by the additional complication.

It is an open question whether the German "Karl" Howitzer of 1942 qualifies as an SP or an auxiliary propelled equipment. On the face of it, it looks like an SP gun, with its enormous tracked carriage, Mercedes Benz V-12 diesel engine and automatic transmission, but in fact these

244
The British wheeled self-propelled 40mm Bofors gun frequently found employment as an infantry support weapon.

245
Duster. For protection of mobile columns, and as a handy assault gun, twin 40mm Bofors guns were mounted on a tank chassis in this American Second World War equipment.

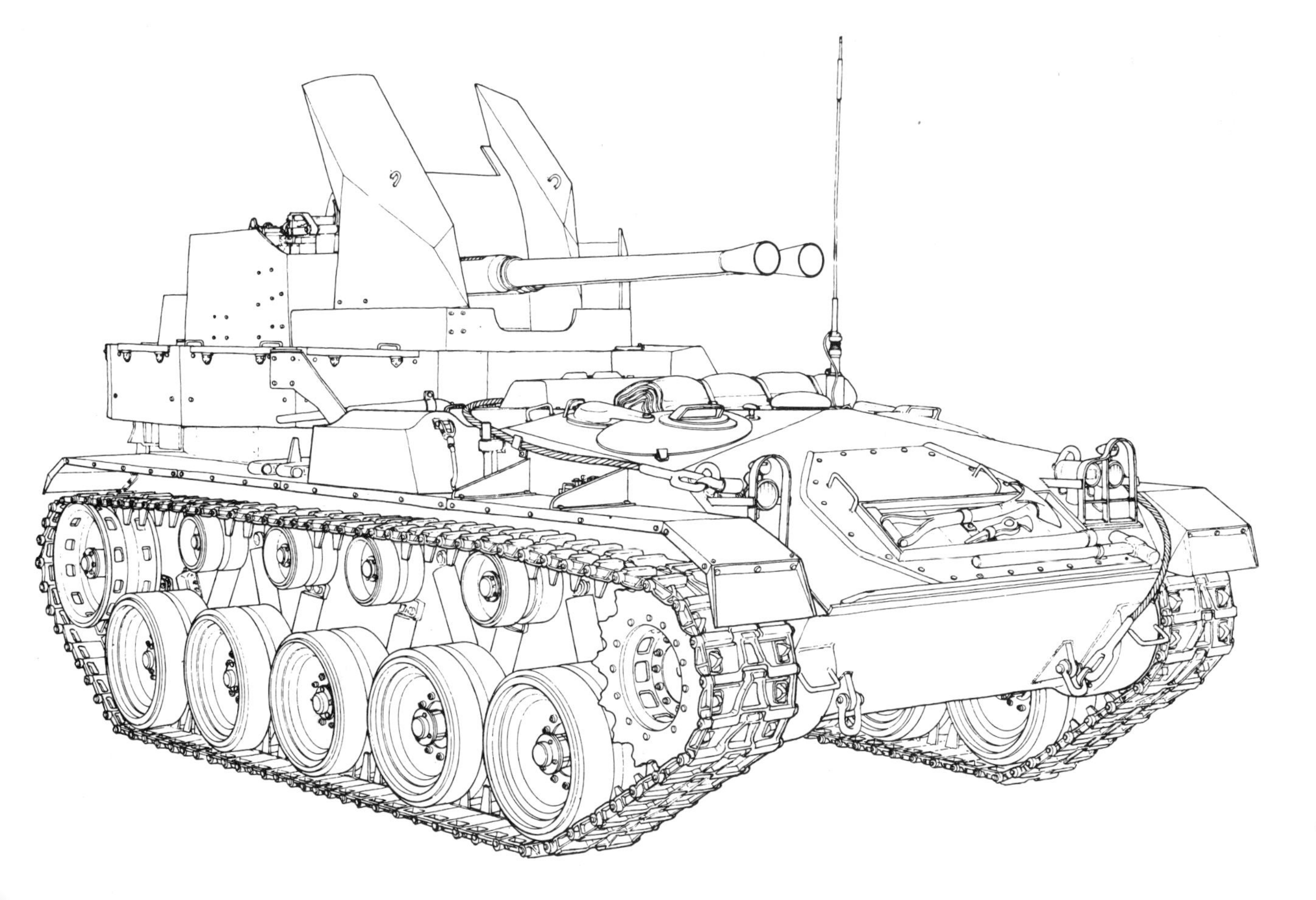

246 **25-Pounder Centurion.** A Post-war British experiment was this 25-pounder field gun mounted into a Centurion chassis, abandoned because it was too much chassis for too little gun.

weapons (241) moved only for short distances under their own power. When large distances were to be covered the barrel, breech ring, top carriage and recoil system were dismantled and loaded onto a special 16 wheel trailer while the carriage was winched aboard a tank transporter.

For even larger moves, into and out of a battle zone, special cantilever girders were bolted to the carriage and the complete equipment suspended between two railroad flatcars for transport by train. Six of these weapons were built commencing in 1939 (241) fitted with 60cm barrels, and were employed as seige guns on the Russian front, notably at Brest Litowsk, Leningrad and Sebastopol. In 1943 some 54cm barrels were produced to suit the carriages, to be interchangeable with the 60cm barrels, but there is little evidence of these having been used in action.

One last German development might be considered here. As discussed elsewhere, the German Army had large numbers of railroad guns, but towards the end of the war Allied bombing had disrupted the German railways to such an extent that difficulties could be foreseen in the movement of these weapons in a combat zone. To enable railroad guns to bypass damaged track, a design was proposed in which the body of the gun could be jacked off its tracks and lowered onto two converted Tiger tank chassis, one at each end. With the tank drivers in communication by telephone the two tanks would then drive off around the damaged section of track, cross country if need be, with the rail tracks being brought along by special transporters. It was considered feasible to use the Tiger tanks to carry the gun to its firing point if need be and allow the gun to be fired from them. It is believed that at least one such piece of equipment was experimentally built and operated, but the idea came too late to be put into production. Undoubtedly this ought to qualify as the biggest SP gun of all time.

247 **5.5 inch Centurion.** Another Centurion-based conversion, the 5.5 inch medium gun was a more practical proposition but development was stopped as being uneconomical, since supplies of the US 155mm Howitzer M44 were readily available and 155mm had been agreed upon as a NATO standard calibre.

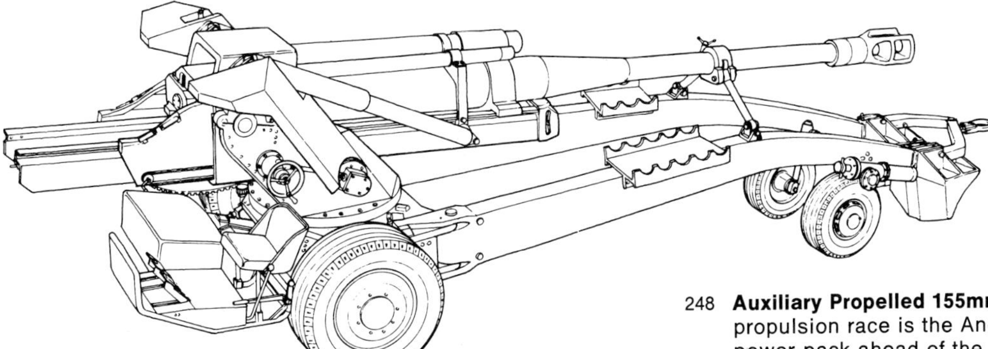

248 **Auxiliary Propelled 155mm Gun.** The latest entrant in the auxiliary propulsion race is the Anglo-German-Italian FH70 which has a power-pack ahead of the axle to give short range mobility.

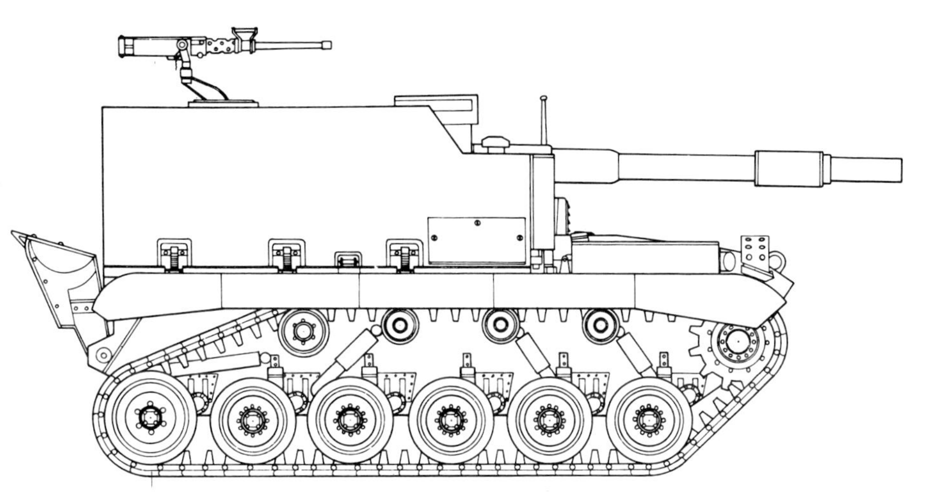

249 **US 155mm Howitzer M44.** The M41 was replaced in the middle 1950s by this design, which incorporated such refinements a power elevation and traverse, electric firing and a much improved chassis.

250 **US 105mm M52.** Latest in the American 105mm SP Howitzer family, this model is turretted for nuclear safety and also for maximum flexibility of fire.

251 **US 8 inch Howitzer M55.** This equipment has the gun, crew and even driver inside a rotating turret which is sealed against nuclear fall-out. Power controls and ramming are provided in order to reduce the number of crew needed.

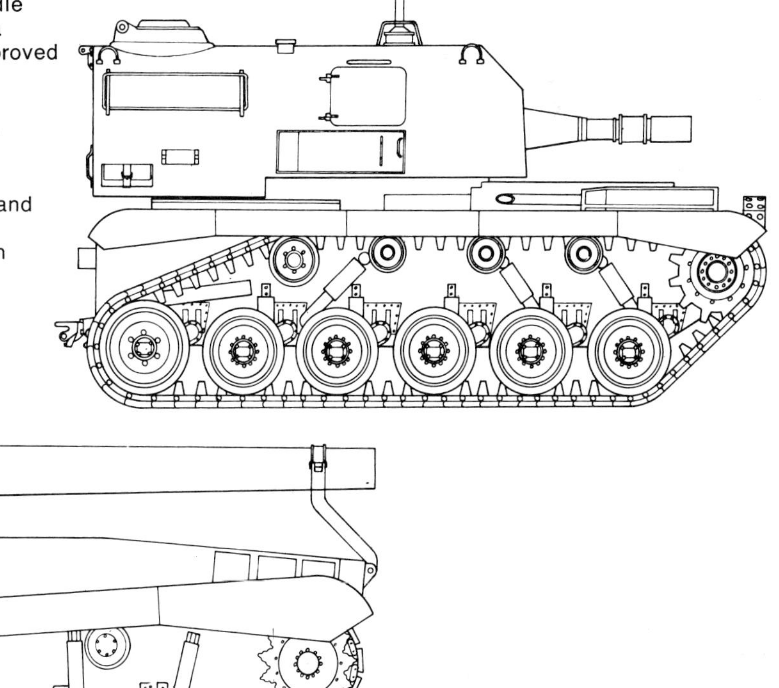

252 **US 155mm M109.** The current general support weapon of the British and US Armies is this 155mm gun, fully powered and with fume extractor and muzzle brake. *Photo: RAC Museum.*

253 **US 175mm M107.** The current long-range divisional weapon of the NATO forces in Europe, this has a surprising turn of speed. The barrel is retracted for travelling, and full power assistance is provided for elevation, traverse and loading.

254 **Abbot.** Britain's current close support weapon, the Abbot 105mm Gun is probably the best SP field gun in existence today.

255 **The Straussler Conversion.** An early attempt at providing auxiliary power for a towed gun, the Straussler was turned down as too cumbersome and too difficult to dig in, as well as having limited cross-country performance. The basis can be recognised as a British 17-pounder anti-tank gun.

256 **Soviet 85mm Gun.** A Soviet auxiliary-propelled weapon, the 85mm gun uses a power pack clamped to the trail to drive through a steerable dolly wheel.

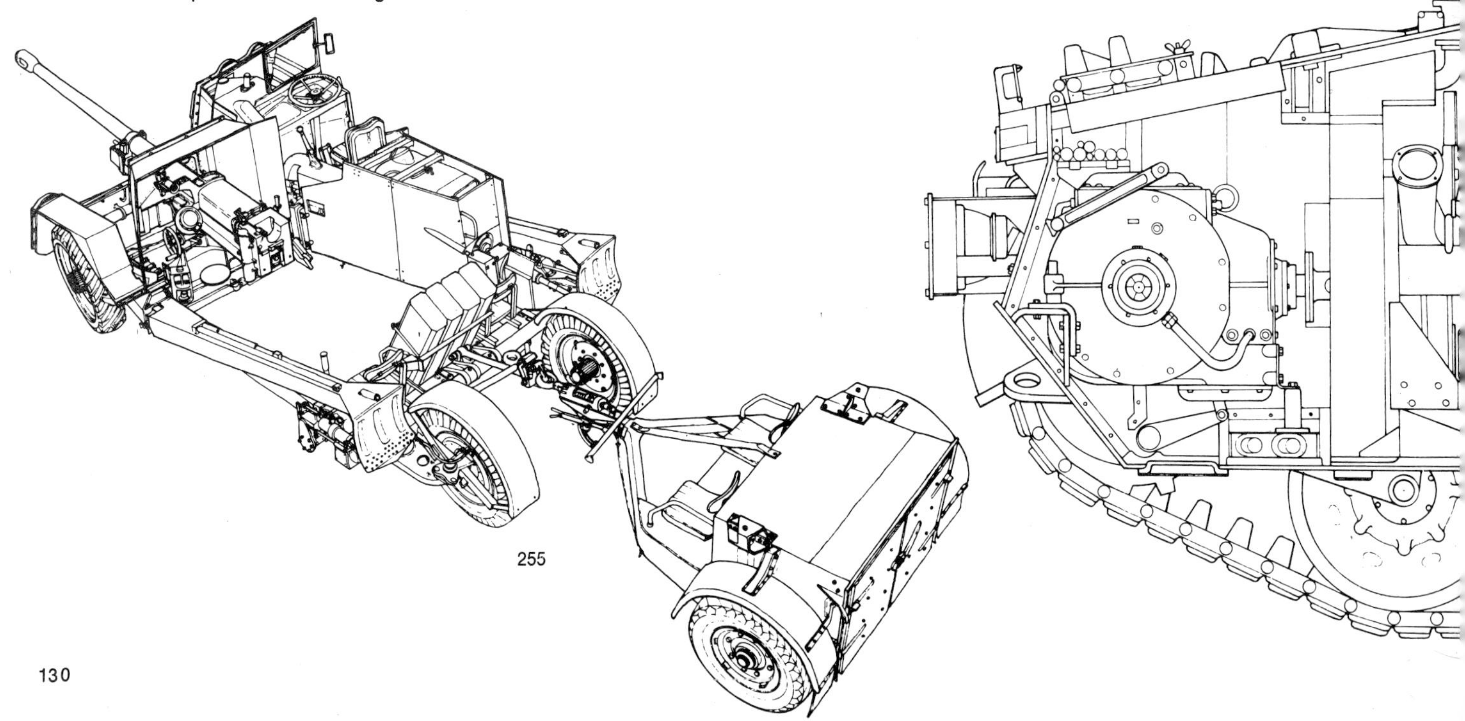

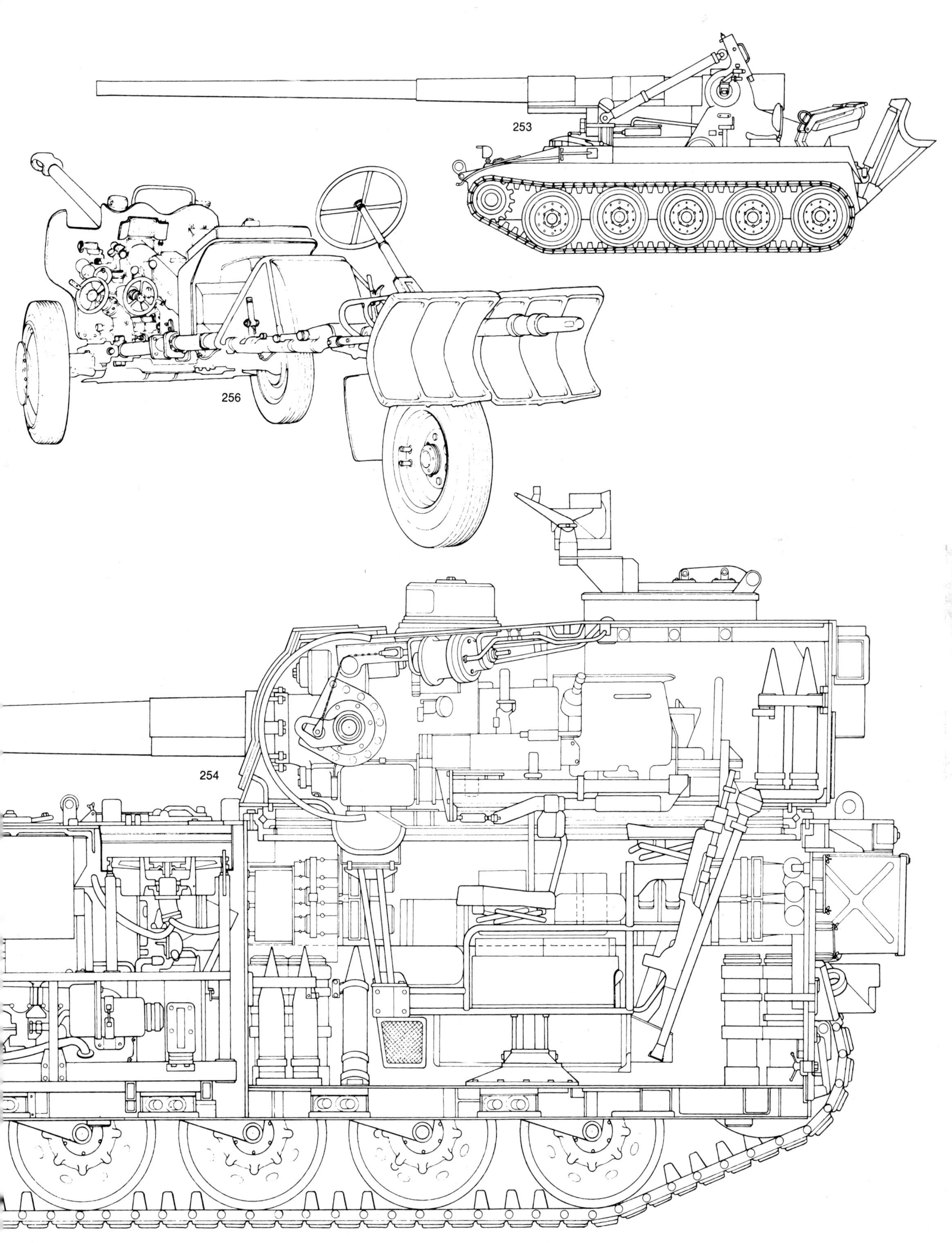
253
256
254

257
Abbot, among the trees, demonstrating its high angle firing ability.
Photo: Vickers.

Recoilless guns

The British 32-pounder and the German PAK 43/41 showed the way that the anti-tank gun was going. It was becoming far too heavy and cumbersome, since heavy armour demanded high velocity to pierce it with kinetic energy ammunition. It was thus necessary to do a radical rethink if weapons of more convenient size were to be produced for field use, and the anti-tank gun of the classic type, with long barrel, powerful charge, and violent recoil, was doomed. The stage was set for the lightweights to take their place. And the first of these was the recoilless gun. As we saw earlier, the hope of doing away with recoil had attracted inventors for many years. The first man to make a successful weapon of war was Commander Davis of the US Navy who, just prior to the First World War, developed the Davis Countershot Gun (259). This worked on a simple enough principle; place the cartridge in between two barrels, place the projectile in one barrel and an equal weight of grease and small shot in the other. Fire the cartridge; the projectile would go forward at a certain velocity, and the countershot would go to the rear at the same velocity. Both barrels would have the same recoil force and both would attempt to recoil. Since the forces were equal they would cancel each other out and the gun would be recoilless. The Davis was made in a number of sizes and 2-, 6- and 12-pounder versions were adopted in small numbers by the British during the First World War as aircraft weapons, since only this system would permit a heavy weapon to be fired from an airplane. They were to be used as anti-Zeppelin and anti-submarine guns, but before they were fully operational the war ended, and they were declared obsolete very rapidly.

It is said that the Russians used some sort of recoilless gun during the 1930s and even took some on their abortive invasion of Finland in 1940, but no details are available on these weapons and it is difficult to decide on what principle they worked. The next major use of recoilless guns was by the German airborne troops who invaded Crete. They were equipped with the 75mm Light Gun (LG2), which fired the conventional 75mm gun shell but carried the Davis gun

258

The Davis Gun. Invented just prior to World War One, the Davis recoilless gun was tried by the British Royal Naval Air Service in an attempt to provide aircraft with a powerful anti-submarine weapon.

259

Davis Gun Principle. The Davis gun achieved recoillessness by firing an equal weight countershot to the rear to balance the recoil due to firing the normal projectile from the front barrel.

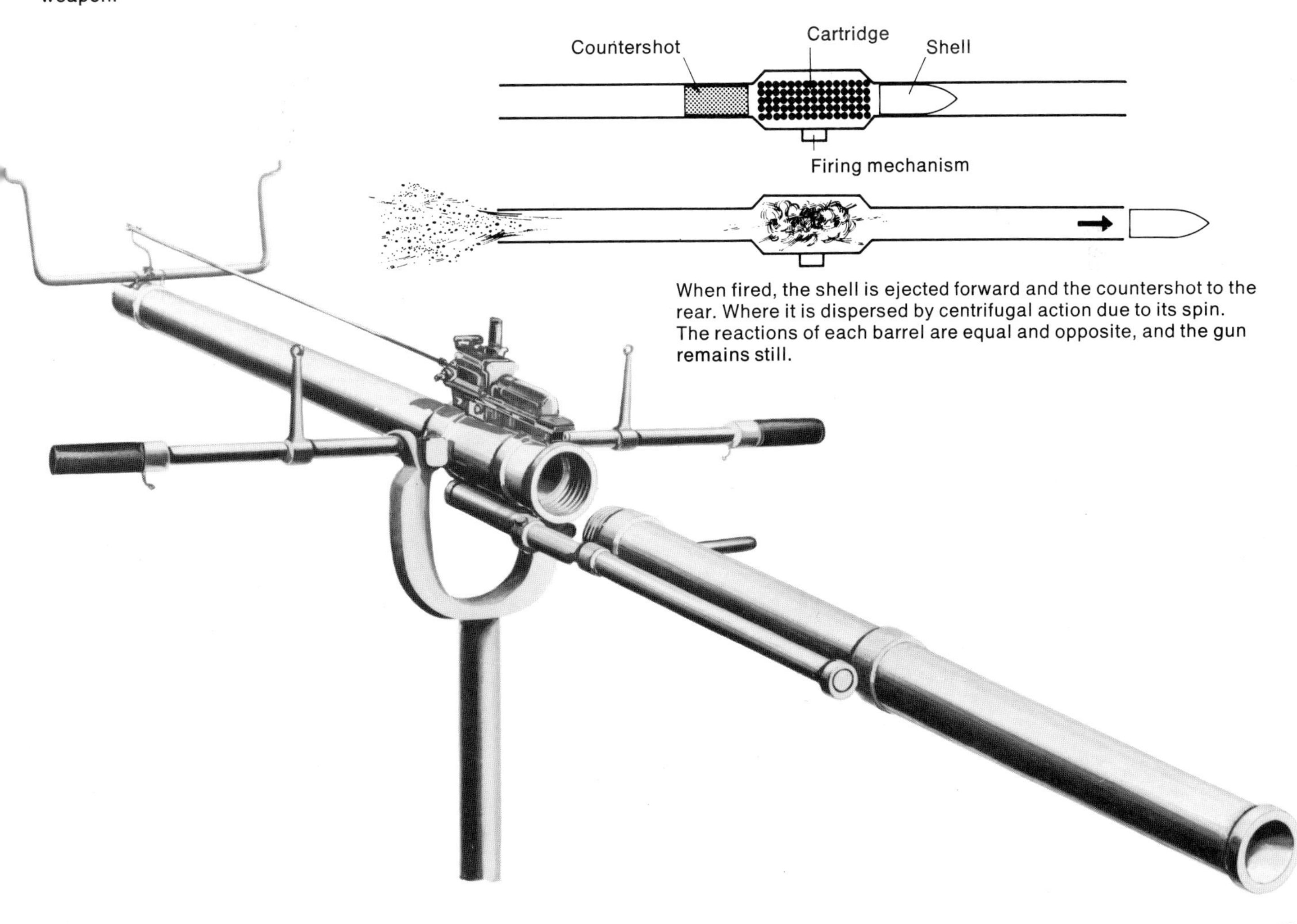

When fired, the shell is ejected forward and the countershot to the rear. Where it is dispersed by centrifugal action due to its spin. The reactions of each barrel are equal and opposite, and the gun remains still.

260 **German 75mm Light Gun.** The German Army unveiled the recoilless gun as a practical proposition during the invasion of Crete. It was later issued to mountain troops and others who required lightweight firepower.

261 **Light Gun Principle.** This shows the internal arrangements of the German recoilless gun. The cartridge case had a blow-out plastic base, and the firing mechanism was suspended in a pod in the centre of the venturi, a location which gave rise to considerable trouble from gas erosion.

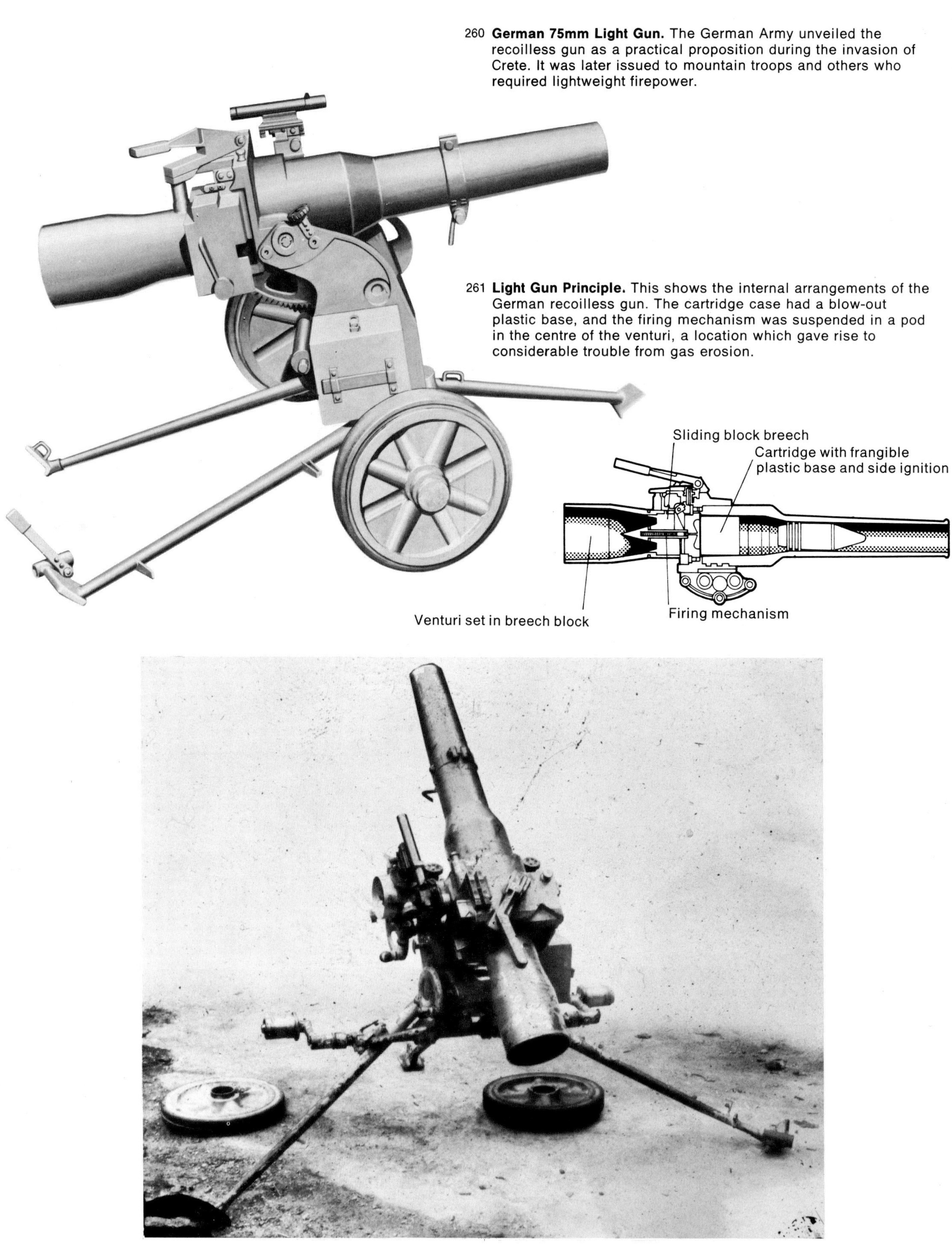

262 German 105mm RCL showing the sliding block breach carrying the venture unit.

to its logical extreme. If it is possible to fire equal weights at equal velocities in opposite directions and achieve recoillessness, then it follows that one can fire half the weight at twice the velocity, a tenth the weight at ten times the velocity, and so on, until one arrives at the point where it is possible to fire a stream of high-velocity gas to the rear and still have the gun without recoil. This is what the German LG2 did; a special cartridge case with a thick plastic disc in the base was used. When the charge exploded, this base was blown out and the propellant gas was allowed to escape through a venturi in the breech block, giving the necessary reaction to the recoil due to the shell passing up the barrel (261). Early versions used a tiny primer in the middle of the base and a firing mechanism in the middle of the venturi, but the gas wash soon ruined the firing mechanism and later designs placed the primer in the side of the case and mounted the firing mechanism over the breech.

105mm and 150mm versions of this gun were later produced, and many other recoilless developments took place with a view to repeating the Davis gun idea and carry a heavy weapon aloft to deal with ships and other heavy targets.

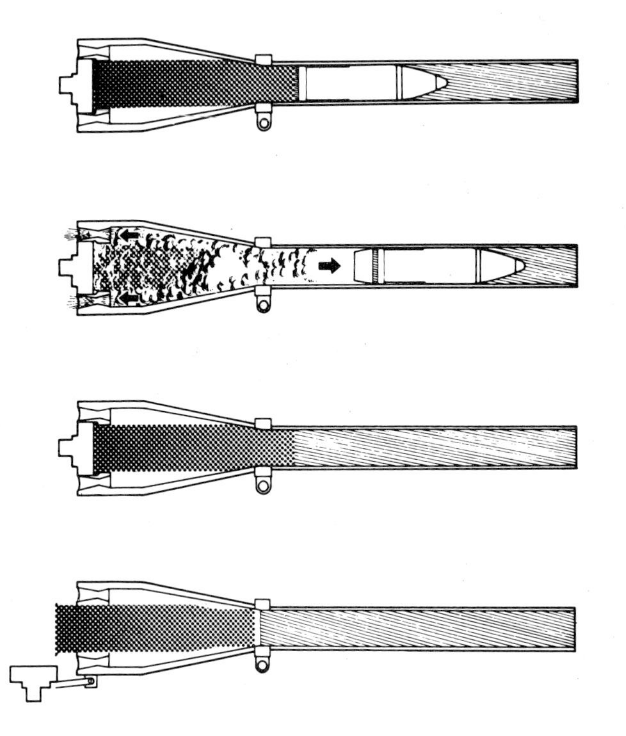

263 **The Kromuskit.** The US recoilless gun in diagrammatic form. The perforated case allowed gas to leak into the annular space and be vented to the rear to give the necessary balance. The shell's driving band was pre-engraved in order to keep pressures low.

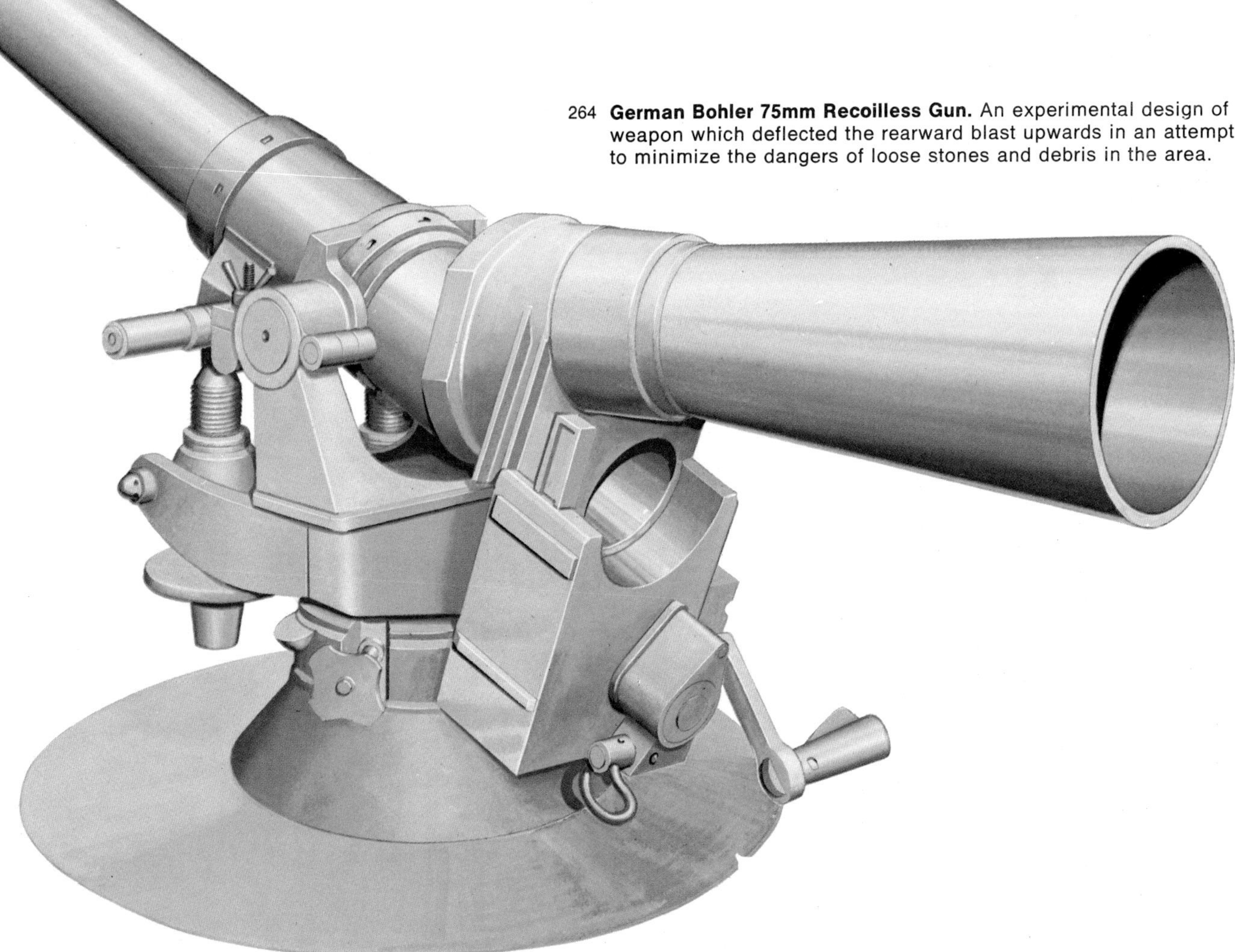

264 **German Bohler 75mm Recoilless Gun.** An experimental design of weapon which deflected the rearward blast upwards in an attempt to minimize the dangers of loose stones and debris in the area.

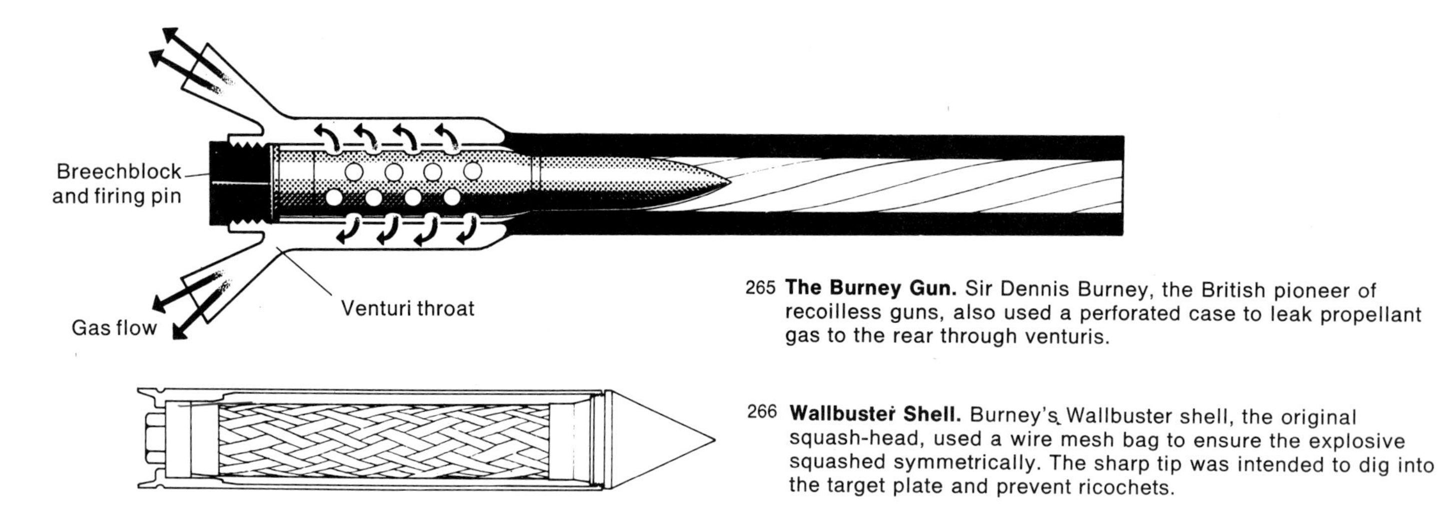

265 **The Burney Gun.** Sir Dennis Burney, the British pioneer of recoilless guns, also used a perforated case to leak propellant gas to the rear through venturis.

266 **Wallbuster Shell.** Burney's Wallbuster shell, the original squash-head, used a wire mesh bag to ensure the explosive squashed symmetrically. The sharp tip was intended to dig into the target plate and prevent ricochets.

In Britain the designs of Sir Denis Burney were rather different. He produced a cartridge case with a number of holes in the side, lined with thin shim brass. The gun chamber was also pierced with holes, and an annular space led back to a number of venturis located around the breech mechanism. When the gun fired the explosion of the cartridge punctured the shim brass and allowed the blast to pass into the space and back through the jets to provide the necessary reaction (265). Originally tried on a four-bore shotgun, it was then tried as a 20mm shoulder gun for infantry, but this was soon dropped in favour of a 3·45 inch shoulder-fired gun (269). Then 3·7 inch, 95mm and 7·2 inch guns were all produced (273). The 3·7 was to be a light anti-tank gun, the 95mm was to be a light field howitzer for airborne troops, and the 7·2 inch was designed simply and solely as a concrete-smashing implement for use in the forthcoming invasion of Europe. As it happened, other weapons were as effective against concrete as the 7·2, so that one was dropped. But the 3·45, 3·7 and 95mm were well received and were intended to be shipped to the Far East, since they looked like being ideal weapons for jungle warfare. The 3·45 inch, for example, was the same calibre as the 25-pounder, fired a formidable Wallbuster shell and could be carried and fired by one man; it was ideal for bringing heavy fire to bear in the confined spaces of a jungle trail, for example. But the sudden collapse of

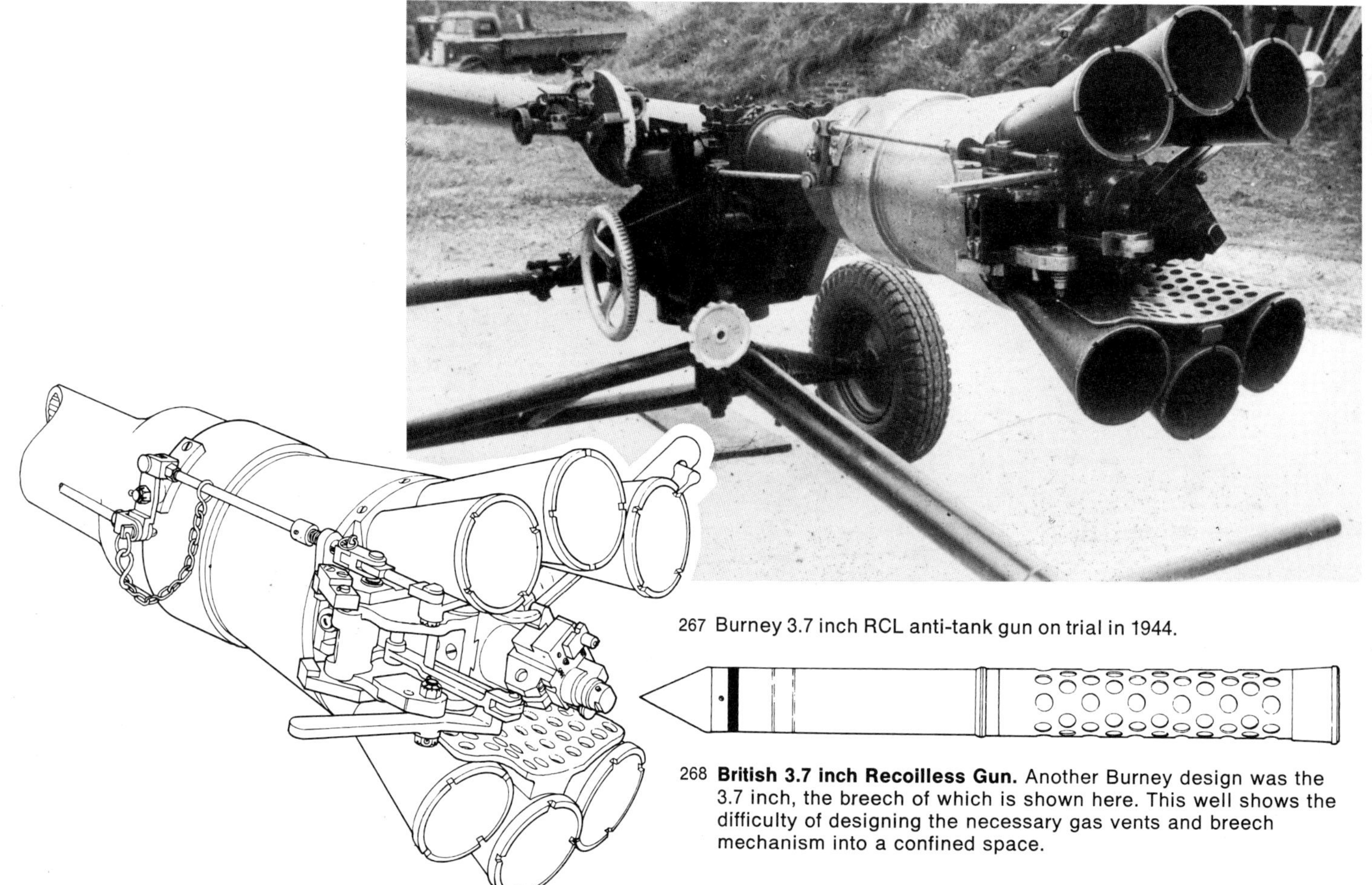

267 Burney 3.7 inch RCL anti-tank gun on trial in 1944.

268 **British 3.7 inch Recoilless Gun.** Another Burney design was the 3.7 inch, the breech of which is shown here. This well shows the difficulty of designing the necessary gas vents and breech mechanism into a confined space.

Japan stopped them before they were shipped, and a small number of 3·45 and 3·7 were issued to selected infantry units to be tried both technically and tactically to see if the recoilless anti-tank idea was worth pursuing.

In the USA similar designs were being developed; the first US recoilless design was a straight copy of the German 105mm Light Gun, and it was well on the way to becoming approved and issued when another design appeared. This was the Kromuskit – named from its inventors, Kroger and Musser, and was very similar to the Burney design, using a perforated cartridge case feeding to rearward-pointing jets. But there was one additional feature of the Kromuskit; the driving band was already engraved with the rifling marks. This meant that the explosion no longer had to force the copper into the rifling and thus the gun developed less pressure and could have a lighter construction. It also meant that you had to devise some method of indexing the cartridge so that the pre-rifled band entered the rifling in proper alignment, but this was done by placing indexing pieces on the cartridge case.

57mm and 75mm Recoilless Rifles (275) were made and tried with great success, and these were adopted as the standard weapons of the US Army, the 105mm falling by the wayside. Numbers were despatched to the South

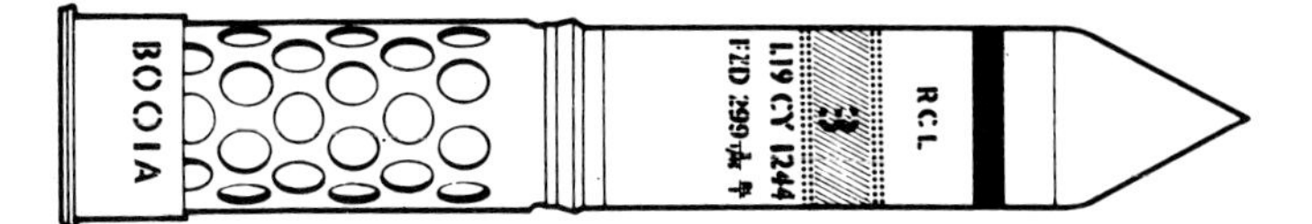

269

British 3.45 inch Recoilless Gun. The first British Burney gun was this, frequently referred to as the 25-pounder shoulder gun since it was of the same calibre. But the round of ammunition was completely different, being fixed and with the perforated cartridge case.

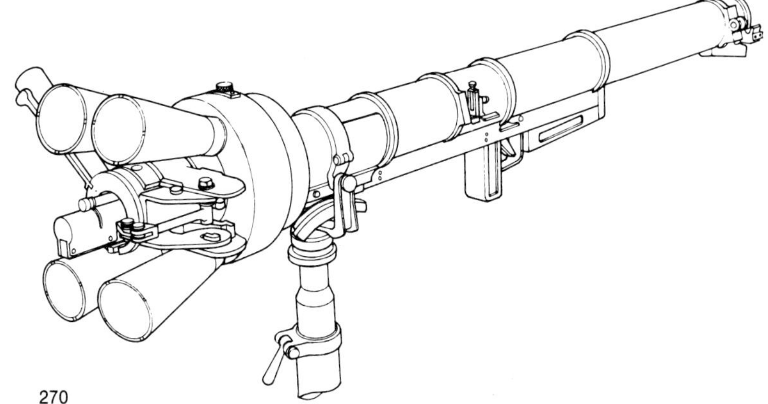

270

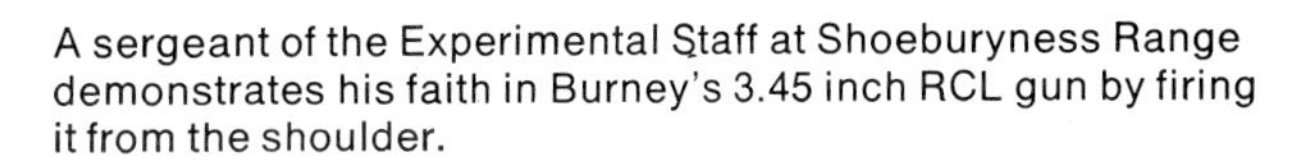

A sergeant of the Experimental Staff at Shoeburyness Range demonstrates his faith in Burney's 3.45 inch RCL gun by firing it from the shoulder.

271

British 4.7 inch Recoilless. This, the last of the Burney designs, was taken over at the war's end by Government design teams and was used as the basis of development of the 120mm BAT.

272

The British 7.2 inch recoilless gun fires, causing as much havoc behind as before.

273

British 7.2 inch Recoilless. A British design intended to provide a lightweight gun with a heavyweight punch for assaulting concrete defences in the invasion of Europe. Since other weapons already in service could do the job just as well, the 7.2 was never adopted.

273

Pacific and were used in the latter weeks of the war, earning themselves a good reputation.

The results of this brief action and the British extended trial was to confirm the adoption of recoilless guns as anti-tank weapons for use by the infantryman. A Burney design of 4·7 inch was under way when the war ended and this was taken over by the British Royal Armament Research and Development Establishment. After some years of redesign and experiment it emerged once more as the 120mm BAT (Battalion Anti-Tank gun) (274), but it is obvious that Burney's principle had been severely modified. The perforated cartridge case was abandoned and a design of plastic-base case adopted. Although apparently the same as the wartime German design, many years of research went into this case design, and it is much improved over the wartime models. Since the introduction of the BAT it has gone through various changes, becoming the CONBAT, the MOBAT and others which did not arrive in service. All these are the basic BAT with simplified mountings, improved sights, the addition of spotting rifles and so forth.

The US also took to their drawing boards and spent some years developing a fresh generation of recoilless guns, although they have stayed with the original Kromuskit design, as seen by the 106mm BAT (276). All these guns are highly effective; firing hollow charges or squash-head ammunition, they can dispose of anything which moves, are light, handy to manoeuvre, and simple to operate. But they retain the drawback that has bedevilled recoilless guns ever since Davis; it doesn't pay to stand behind them. The sheet of flame which comes out of the rear end of a recoilless gun is awesome to behold, and renders the

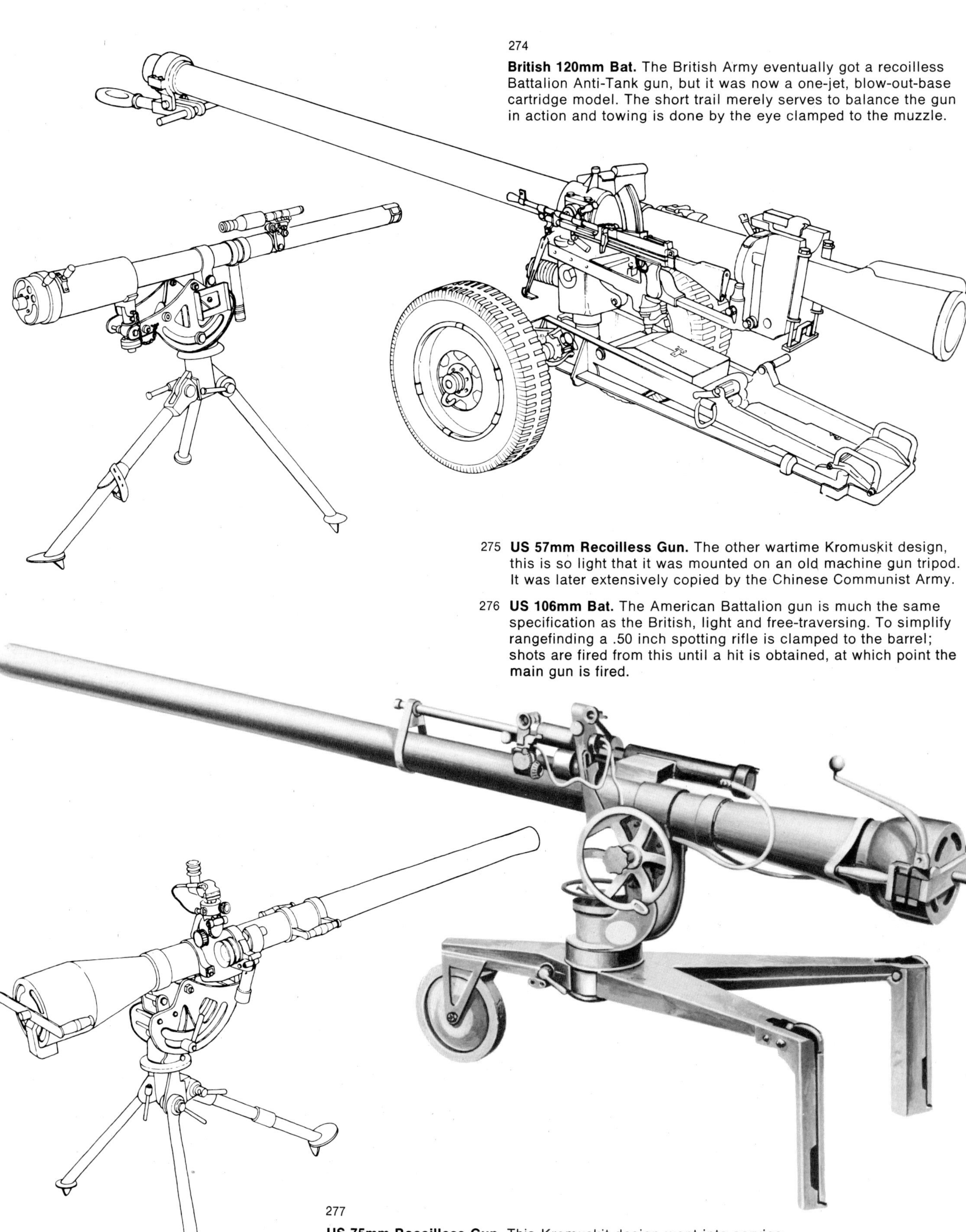

274

British 120mm Bat. The British Army eventually got a recoilless Battalion Anti-Tank gun, but it was now a one-jet, blow-out-base cartridge model. The short trail merely serves to balance the gun in action and towing is done by the eye clamped to the muzzle.

275 **US 57mm Recoilless Gun.** The other wartime Kromuskit design, this is so light that it was mounted on an old machine gun tripod. It was later extensively copied by the Chinese Communist Army.

276 **US 106mm Bat.** The American Battalion gun is much the same specification as the British, light and free-traversing. To simplify rangefinding a .50 inch spotting rifle is clamped to the barrel; shots are fired from this until a hit is obtained, at which point the main gun is fired.

277

US 75mm Recoilless Gun. This Kromuskit design went into service in time to see action in the South Pacific in 1945 and was later extensively used in Korea.

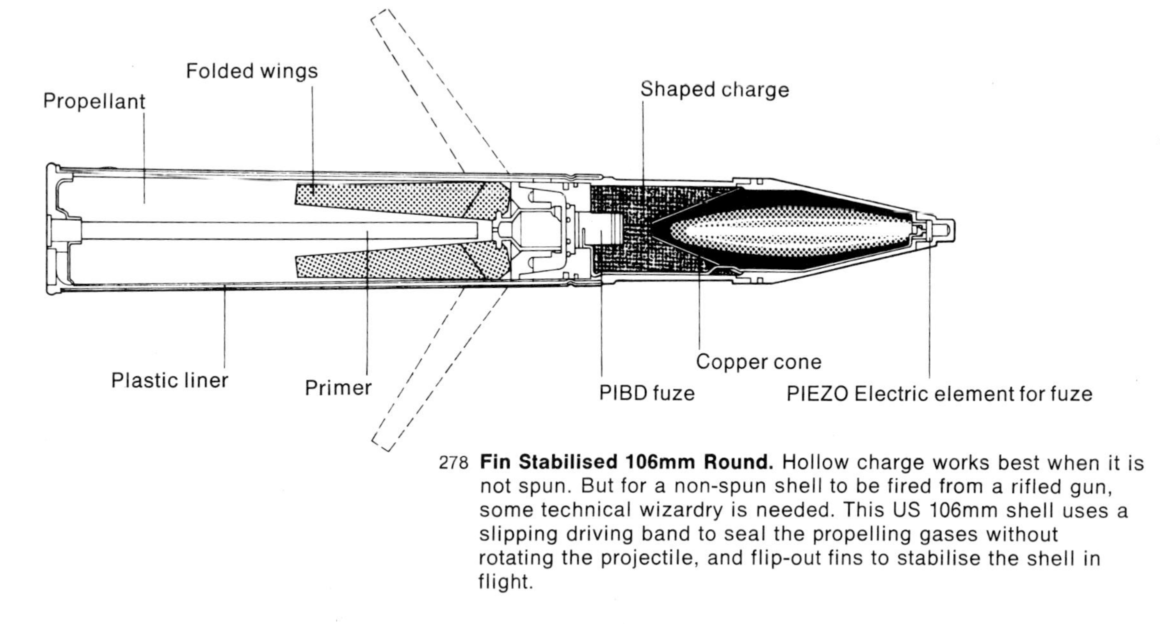

278 **Fin Stabilised 106mm Round.** Hollow charge works best when it is not spun. But for a non-spun shell to be fired from a rifled gun, some technical wizardry is needed. This US 106mm shell uses a slipping driving band to seal the propelling gases without rotating the projectile, and flip-out fins to stabilise the shell in flight.

279 105mm Recoilless gun, mounted on a jeep, being operated by a French Army detachment.

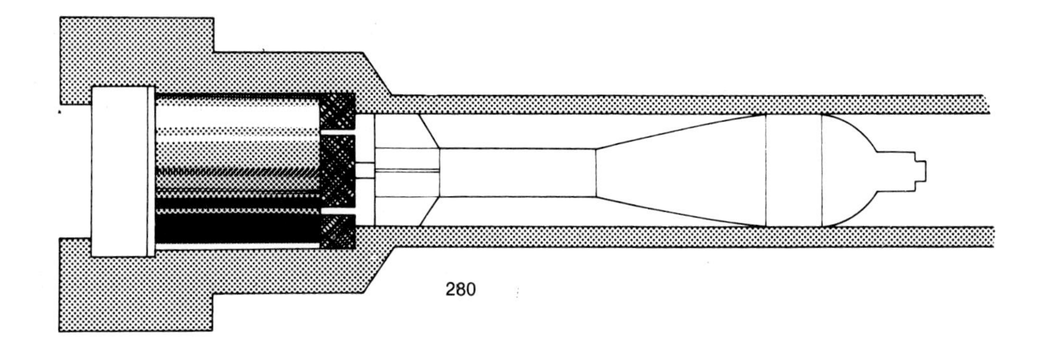

weapon difficult to site and practically impossible to conceal once fire has been opened. But if you offer the infantry the 32-pounder in exchange, I feel they would prefer to keep their BAT, flame an'all.

The rearward blast and flame of course, highlights one of the basic defects of the recoilless gun – its appetite for propellant. Modern weapons are rather more efficient, but I remember that the rule of thumb in the days of the Burney gun was that four-fifths of the propellant went out of the back, and one-fifth actually did work in pushing the shell out of the front. It was this prodigality with explosive which led the German Army to confront the gun designers in 1944 with a demand for a weapon as light as a recoilless gun but which used less propellant, was still accurate enough to put every shot into a one-metre square at 500 metres' range, and was capable of defeating any enemy tank. Most designers didn't want to know, but Rheinmetall took it up and produced in very short order a gun which introduced a completely new ballistic concept – the High-Low Pressure Gun (280). It was a smoothbore weapon, of very light construction, using a hollow charge fin-stabilised projectile which had distinct affinities with a mortar bomb. The cartridge case was that from the 105mm Howitzer, and it was fitted at its mouth with a heavy steel plate pierced with a number of venturi-like holes. This plate carried a spigot, to which the projectile was attached by a shearable joint. The gun chamber had a distinct shoulder at the front, and when the round was loaded, the projectile sat in the barrel, while the heavy steel venturi plate was butted firmly against the shoulder. On firing, the propellant exploded in the cartridge case, generating a high pressure – something in the order of eight or nine tons per square inch. This was then bled through the venturi into the space behind the projectile, so that the "chamber" pressure behind the bomb was only about two or three tons. This sheared the joint and the bomb was launched from the gun barrel.

Thus the gun chamber had to be robust, but the rest of the barrel could be thin and light, with only a low pressure to worry about. The resultant gun, the 8cm Panzerabwehrwerfer 600 (281) was produced in small numbers before the war ended and put a stop to its short life. But the principle was brilliantly effective, and caused a lot of ballisticians to look closely at it when captured specimens became available. Strangely, little use has been made of this principle in the ensuing years; perhaps the only weapon in which this principle has been admittedly used is the US Army's M79 Grenade Launcher, in which the high-pressure chamber actually forms part of the cartridge case. Other weapons have doubtless been developed but nobody's talking about them at the moment.

280

The High-Low Pressure System. The projectile, a hollow charge bomb, is separated from the propellant by a thick plate pierced with holes. When the cartridge is fired, the propellant developes a high pressure in the chamber which is bled through the holes to produce a low-pressure push to propel the projectile. This allows the barrel to be made much thinner and thus save weight.

281

German Panzer Abwehr Werfer 600. This weapon was a lightweight gun with a powerful punch, due to using the High-Low Pressure principle. It was an attempt to produce a weapon as light as a recoilless gun but without the appetite for propellant powder.

Fuzes

While the gun is the outward and visible symbol of the Gunner, the fact remains that the weapon of the artillery is the projectile; the gun is really no more than the final link in the transport system which moves the shell from the factory to the enemy. And unless the shell functions in the right place, at the right time and in the right fashion, everybody's efforts in making, transporting and firing the thing have been wasted. Carrying this line of reasoning to its ultimate, one is driven to the conclusion that the most important part of the gunner's armoury is the fuze which initiates the shell. In earlier pages we have briefly mentioned fuzes, generally in connection with some particular application, but by way of reinforcing their importance this section shows a few of the hundreds of different designs of fuze which have appeared over the years.

Originally the fuze was no more than a wooden plug carrying a central core of finely milled gunpowder which would burn at a more or less regular rate. The outer surface was marked off in units of time, and the fuze was cut off to a suitable length before insertion into the shell. Ignition was originally by the gunner, as already described, later by flash-over from the propelling charge. (It is worth bearing in mind, in this connection, that development of a time fuze, however rudimentary, had to await the development of a means of measuring time with some degree of accuracy. An early method of testing burning time was to recite the Creed during the burning of a test fuze, a system which was no doubt good for the spiritual welfare of the gunners but which could hardly be considered accurate.)

The first major step came in the middle 1800s when Colonel Boxer, Superintendent of the Royal Laboratory at Woolwich Arsenal, developed his wooden time fuze. While resembling the earlier models, it was much different internally and allowed a greater degree of accuracy in cutting. (282) The fuze had two separate fillings of gunpowder, and a large number of holes distributed radially around the outside. These holes were covered with a waxed or varnished paper cover which was marked with dots corresponding with the holes, each representing a given time of burning. To use the fuze the gunner merely selected the dot giving him the desired time and drove a pointed awl into the fuze, opening up a hole between the inner and outer powder fillings. When the gun was fired, the flash-over would light the central filling which would burn down regularly until it met the hole bored by the gunner. Ignition would then pass through the hole and the outer filling of powder, a much faster-burning composition, would flash down to the bottom of the fuze and ignite the shell's contents.

This was a considerable advance over its predecessors, being more accurate and easier to set, and Boxer soon set up mass-production facilities at Woolwich to turn these fuzes out by the hundreds of thousands. Unfortunately they were found to be unreliable in service, and a long investigation took place to discover why. It was eventually found that the wood-turning machines used to make the fuze bodies were kept amply lubricated, and some of this oil was finding its way into the wood of the fuzes where eventually it affected the powder filling and caused irregular burning; making the turners operate their machines dry removed the problem.

While the Boxer design solved the time fuze problem of the day, an impact fuze was also required, and this proved slightly harder. Many and varied were the designs produced, and the problems the designers faced and overcame are much the same as the present-day designer has to deal with. Some-body once said that a fuze is a combination lock which can only be unlocked by firing it from a rifled gun. The conditions the fuze has to put up with are strenuous; it has to withstand vast acceleration when fired, high rates of spin in flight, often violent blows from worn guns, and eventually has to operate flawlessly when driven at high velocity into a hard target. To survive all this mechanisms have been designed which take advantage of these conditions to provide safety locks so that the fuze is perfectly safe when being transported and loaded, but automatically cancels out all the safety features during flight so that it is ready to function when it reaches the target.

However, safety was not the principal concern of the early designers; their shells were only filled with gunpowder and a bore premature with one of these was a nuisance but not fatal. Their main concern was reliable operation at the target, and they achieved this with spartan simplicity in some cases. Consider the British Fuze Direct Action No. 3 (286). Here a metal disc holds a needle suspended above a detonator. The disc is cun-

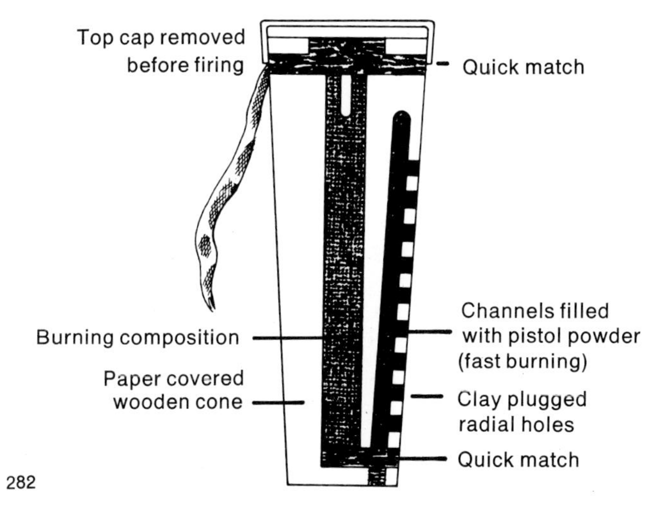

Boxer's Wood Time Fuze. One of the earliest British time fuzes, for smoothbore artillery, this wooden plug carried a filling of gunpowder. When pierced by a sharp awl, the burning powder would flash across from the time channel to the main filling and thus ignite the shell's contents.

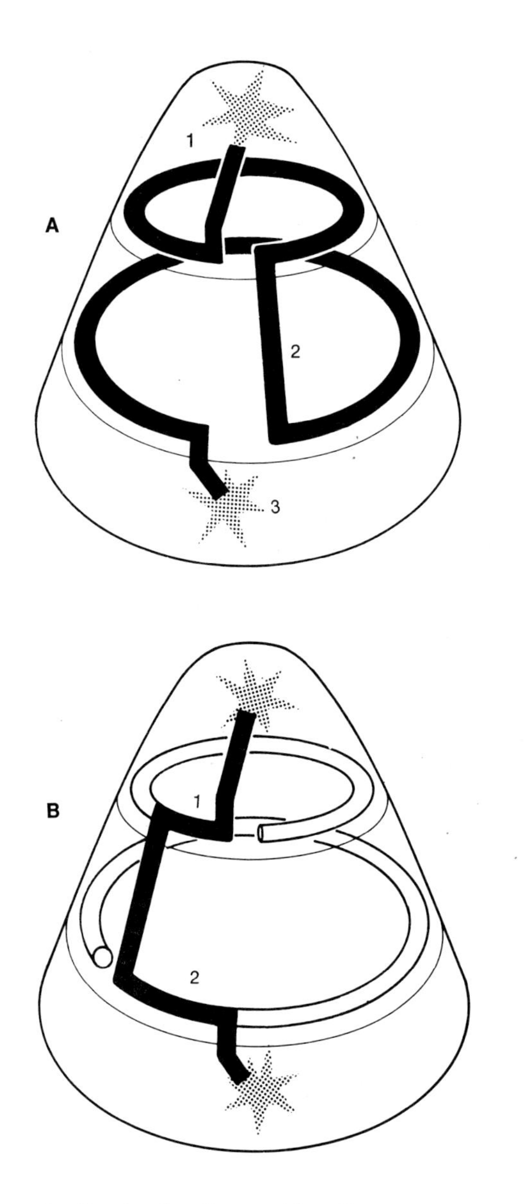

283

Operation of a Combustion Pattern Time Fuze

When the gun fires, the sudden acceleration causes the time detonator to ignite the gunpowder in the top time ring (1). The flame burns round the ring until it reaches a port leading down to the bottom time ring (2). This in turn ignites and burns round in the opposite direction until it reaches a port leading to the magazine, at which point the flame enters the magazine and explodes the powder therein (3) and this in its turn initiates the filling of the shell.

To alter the time of operation between firing the gun and initiating the shell, the bottom time ring is revolved. This displaces the connecting port to a fresh position below the top time ring (1) and also alters the distance from the start of the lower ring's gunpowder train to the magazine (2). Thus when ignited the upper ring burns only a short time before flashing down to ignite the lower ring, which in turn burns for a short time before igniting the magazine.

ningly designed so that it has sufficient rigidity to withstand acceleration and air pressure during flight but will deform readily on impact and allow the detonator to be struck by the needle. The flash from the detonator then passes to a gunpowder filling in the fuze's "magazine", from whence the flash initiated the shell. The fuze was provided with a removable cap which protected the disc during handling and loading, and the disc was also recessed for additional safety. On impact with the ground, earth would be scooped into the recess and this would ensure the disc being depressed to fire the detonator.

This design, of course, pre-supposes that the shell is going to land nose first, a virtual certainty with a spun projectile but an extremely unlikely event with a spherical shell. For this type of projectile it was necessary to produce an "all-ways" fuze which was guaranteed to function irrespective of the angle at which it hit the ground. The Pettman General Service Fuze (285) is a very early example of this type. The large ball is coated with a sensitive detonating composition and held securely in place by wedges at top and bottom. On firing, the sudden acceleration produces "set-back" of the loose components, and the lead cup sets back, shearing the copper supporting wire and freeing the ball. On impact the ball is flung violently in any direction, depending upon the attitude of the fuze, and friction between the detonating composition and the interior walls of the fuze body causes flash; this passes to the gunpowder magazine and the shell is thus initiated. When the spherical shell was supplanted by the spun projectile, the Pettman fuze fell into disuse, but the principle was revived during the First World War for early trench mortar fuzes – since the stability of the bombs was questionable – and revived again in the Second World War for hand grenade fuzes.

When high explosive shells appeared on the scene it was found that safety now assumed a dominant role in fuze design. A faulty detonator, functioning on acceleration, could detonate the high explosive filling, wreck the gun and kill the gunners, so that it was imperative that this type of accident be prevented. The generally-accepted solution to this is to interpose a mechanical barrier between the detonator and the rest of the fuze, a barrier which can only be opened by the various forces at work when the gun is fired. The most common system is to use a centrifugal shutter plate, a solid metal barrier which is held closed by a spring and which can only swing aside and leave free passage for the detonating train when the shell is spinning at its designed rate. The British Fuze Number 45 is an early example of a shuttered fuze; introduced in 1914, and looking remarkably like Boxer's time fuze in outline, it was so successful and reliable that it remained in service with coast artillery shells until 1956. It will have been noticed that these early fuzes made no concessions to streamlining or airflow, being simply designed to work without much regard to what they looked like. In later years when aerodynamics were better understood, it was found that correct shaping of the fuze could add considerably to the range achieved by the shell, and it became normal practise to form the fuze as an extension of the shell's contour at the nose. This, to some degree, made the work of the designer more difficult, since he now had to fit all his safety devices and arming mechanisms into a severely restricted space, and modern fuzes are

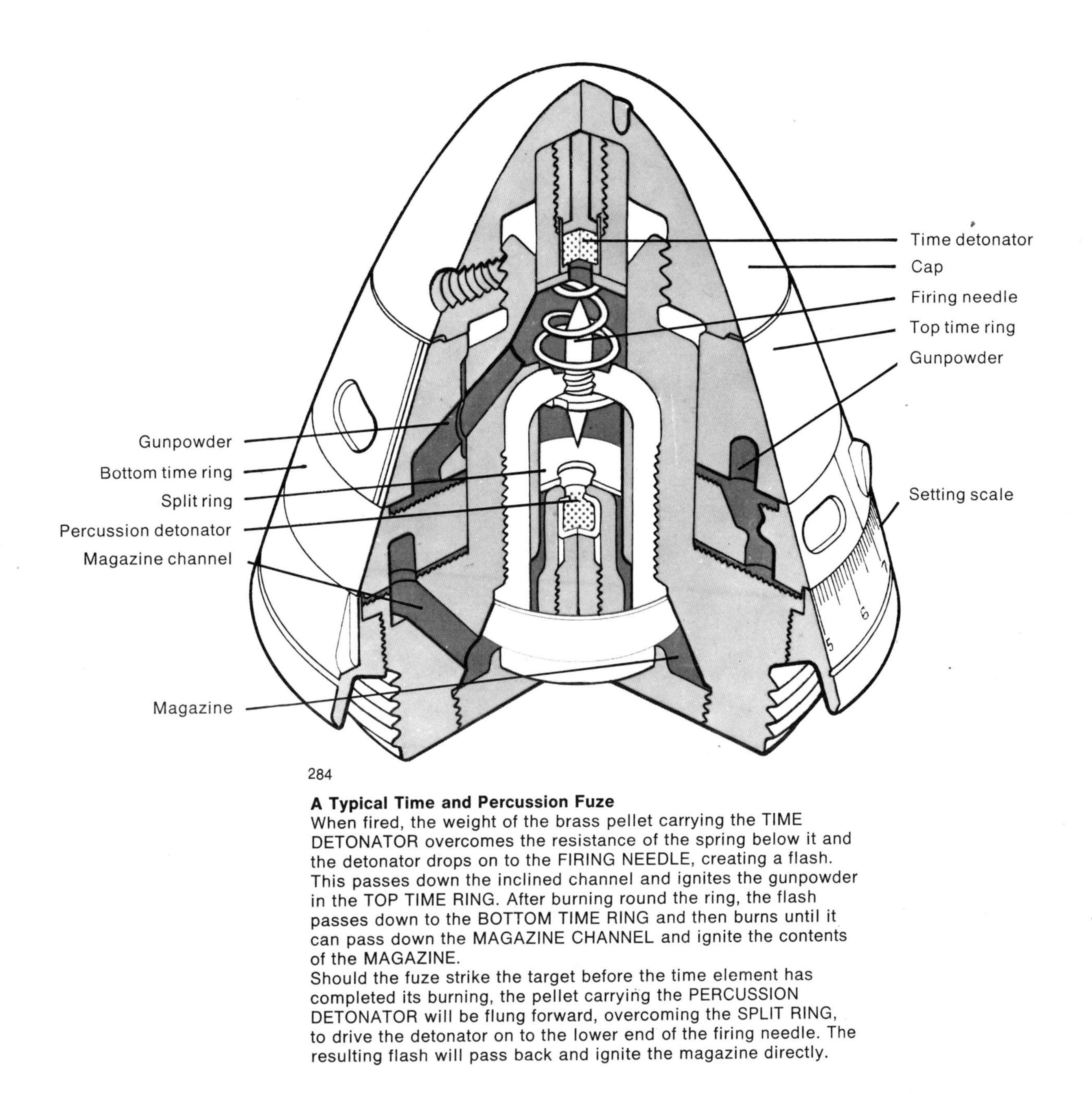

284

A Typical Time and Percussion Fuze
When fired, the weight of the brass pellet carrying the TIME DETONATOR overcomes the resistance of the spring below it and the detonator drops on to the FIRING NEEDLE, creating a flash. This passes down the inclined channel and ignites the gunpowder in the TOP TIME RING. After burning round the ring, the flash passes down to the BOTTOM TIME RING and then burns until it can pass down the MAGAZINE CHANNEL and ignite the contents of the MAGAZINE.
Should the fuze strike the target before the time element has completed its burning, the pellet carrying the PERCUSSION DETONATOR will be flung forward, overcoming the SPLIT RING, to drive the detonator on to the lower end of the firing needle. The resulting flash will pass back and ignite the magazine directly.

filled to the maximum with mechanism. Now it is customary not only to have two or even three independent safety devices, but the users demand such refinements as adjustable or optional delay mechanisms so that the shell can function instantly on impact or a few micro-seconds after impact so as to be able to penetrate light cover before detonation. In anti-aircraft fuzes it is necessary to add a self-destruction feature so that the shell, should it miss its target, is detonated in the sky and does not fall back to earth in an armed condition where it might do more damage than the raider might have, had he been left alone.

When the rifled gun and the spun shell took over, Boxer's time fuze had to be modified. Driving bands did not allow the flash-over to ignite the fuze, and the first change was the addition of a sensitive detonator and a needle suspended on a shear wire at the nose of the fuze. This allowed set-back to drive the needle into the detonator and thus ignite the powder channel, after which the action was the same as before. But the rifled gun also provided longer ranges and thus longer times of flight, and new designs of fuze were needed to provide longer burning times. The radial time fuze provided this facility (284). In this class of fuze the powder train is coiled within the fuze body to obtain the maximum length of powder. In order to increase the burning time still more, additional lengths can be coiled up in a series of metal "time rings"; the usual number is two, but examples with up to five rings have been made. Somewhere in the fuze body a detonator is suspended over a needle, and from this a channel runs to the beginning of the top time ring. This top ring is fixed relative to the fuze body, while the lower ring in the example shown is capable of being turned. The lower ring also carries a channel, from the powder filling to the top surface of the ring; the powder filling is on the underside of the ring and rides on a platform on the fuze body, and in this platform another channel leads to the gun-powder-filled magazine of the fuze.

When the gunner wishes to use the fuze he sets it by rotating the lower ring; we will assume he sets it for maximum time of operation (283 A). On firing, the detonator sets back onto its needle and the subsequent flash passes up the channel and ignites the beginning of the top time ring filling. The filling burns round at its pre-determined rate until it is completely burned, at which point it passes the flash down the channel in the lower time ring and thus ignites the filling. This burns, in the opposite direction, until it reaches the channel in the fuze body leading to the magazine, when the flash passes down and fires the contents of the magazine, initiating the shell. If the fuze be set for a lesser time (283 B) by turning the lower ring, it will be seen from the drawing that this moves the connecting channel in the lower ring closer to the start of the top ring, so that less of the top ring is burned before flash passes to the lower ring. Similarly, the start of the lower ring has moved across the magazine channel so that less of the lower ring burns before the magazine is ignited. The exact balance between these two functions is determined by the designer, and the gunner's task is merely to set the lower ring according to a scale engraved on the outer surface. In the case of fuzes with more

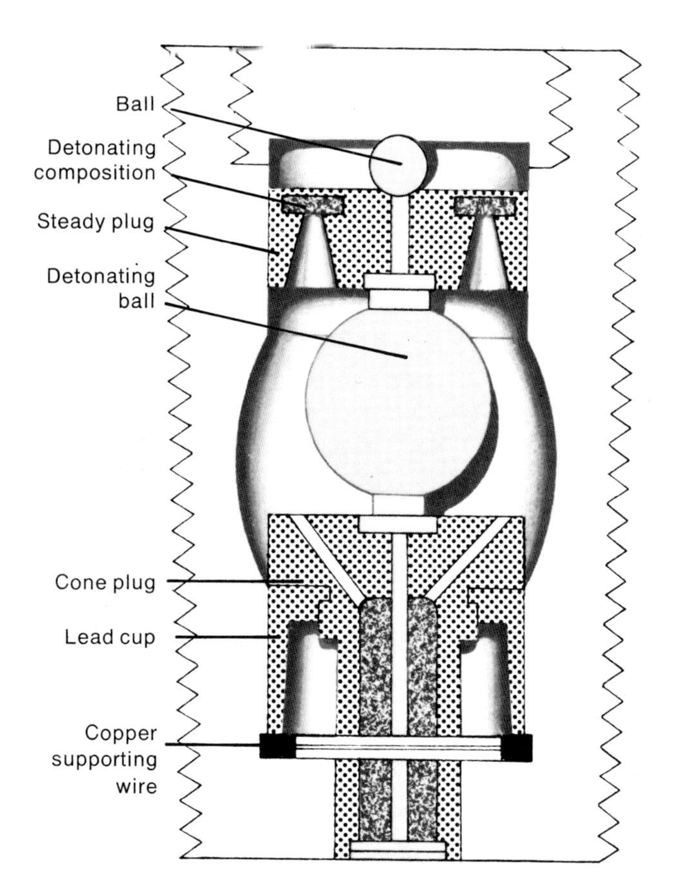

285 **Pettman's Fuze.** One of the earliest 'all-ways' fuzes; on impact at any angle the ball would move. Since it was covered with detonating composition this movement would cause ignition which would then be picked up and amplified by the other components until it ignited the shell filling.

286 **Fuze Direct Action No. 3.** An early and simple fuze for rifled shell. With the shell now guaranteed to land nose-first, this sort of design could be made to work.

287 **Fuze, Time No. 24** Introduced in 1887, this fuze used centrifugal force to throw a detonator carrier against a needle to ignite the powder train. It was set by loosening the top nut, rotating the time ring, and re-clamping the top nut by a spanner.

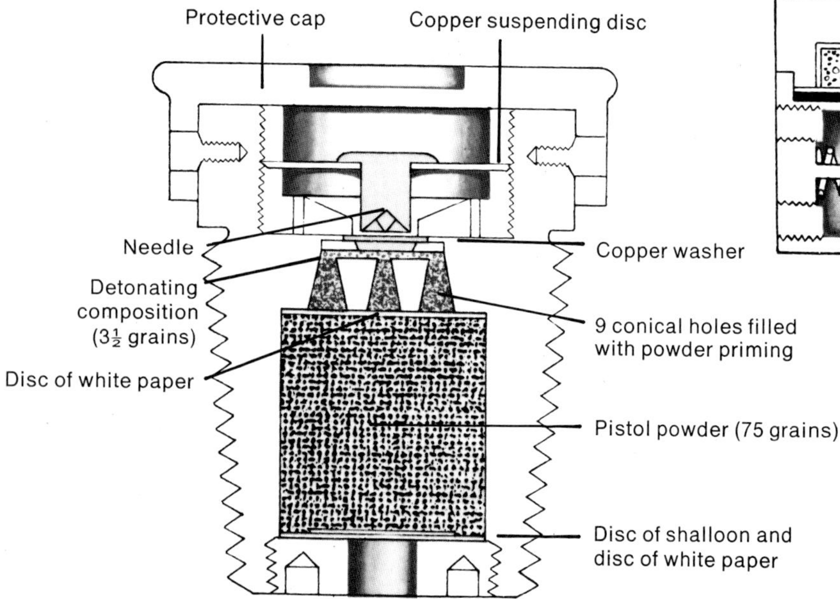

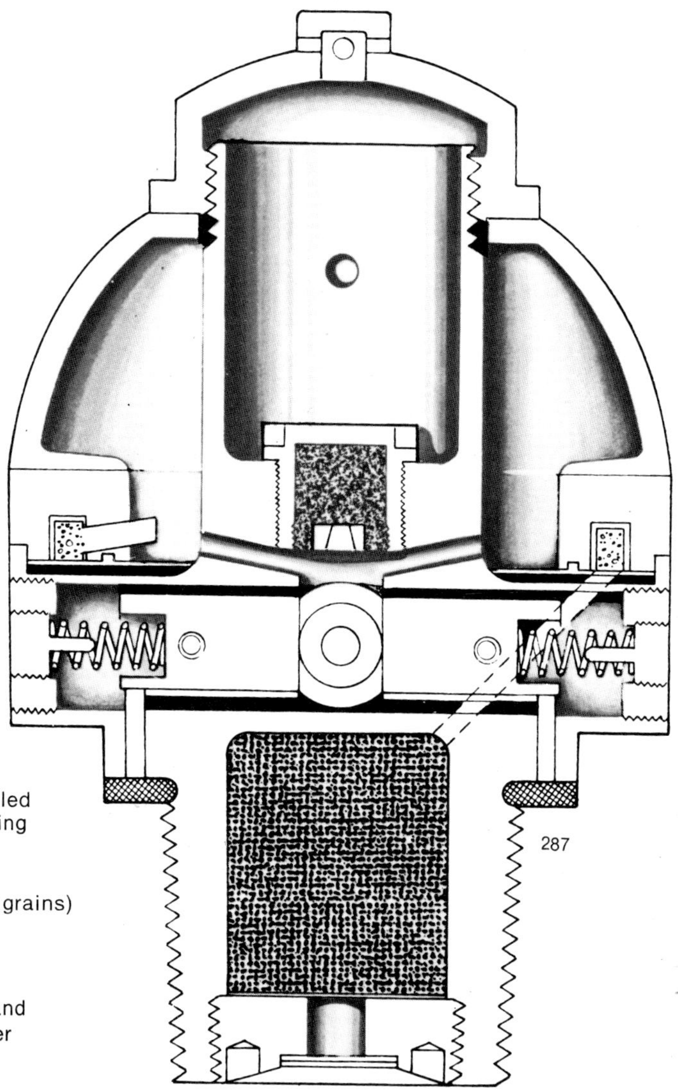

rings, these are arranged so that alternate rings move and are still; thus in the case of a three-ring fuze, the top and bottom rings would be moveable, the central ring fixed. It will also be noted that there are no safety devices with this type of fuze; this is because they are generally used with base ejection or other carrier shell in which a bore premature would be of minor consequence. Where it is necessary to fit them to high explosive shells a separate shutter and detonator unit, called a "gaine", is fitted beneath the fuze, giving both a safety measure and also a method of turning the fuze's flash into a detonation wave suitable for initiating the high explosive filling.
The powder filled fuze has a number of limitations; the powder's rate of burning is sensitive to atmospheric pressure, humidity, temperature, and the rate of spinning of the shell, and abstruse calculations have to be made before firing in order to make due allowances for all these factors. In view of these defects, the Germans began, early in the 1900s, to develop a mechanical fuze which would be independent of all these disturbing elements. Mechanical fuzes had been tried before; the most promising of the 19th century efforts was the 'distance' fuze in which a weight or air-vane was held stationary while the fuze and shell revolved around it. Since the curvature of the gun's rifling was known, it was a matter of simple arithmetic to calculate how many revolutions of the fuze equalled one yard of range, and the mechanism could be suitably calibrated so that when set for a given range it would literally count out the revolutions of the shell until that range had been reached and then detonate. Unfortunately the promise was better than the achievement; Britain fooled around with Thompson's Distance Fuze from 1882 until 1921 without much success; while the principle was sound enough, the acceleration forces usually jammed the mechanism and made it lose count somewhere along the line. In the early low velocity guns it showed promise, but as gun pressures and velocities crept up the distance fuze was out-distanced.
The German idea, later adopted world-wide, was to simply adopt a normal clock mechanism, driven by a spring. The spring was pre-wound in the factory before the fuze was assembled ,and the mechanism was released to run when the shell accelerated. This system, some detractors alleged, meant that the spring could lose tension over years of storage and thus the fuze time would be unreliable, and to defeat this problem a centrifugally driven fuze was developed, using a clock-type regulator with the motive power provided by two heavy weights swinging outwards under the spin and driving, through rack and gears, the regulator. The Germans were notable in employing both types of mechanism impartially during the World War Two, but it is significant that the Americans, long the principal employers of the centrifugal fuze, have recently adopted spring-driven fuzes.

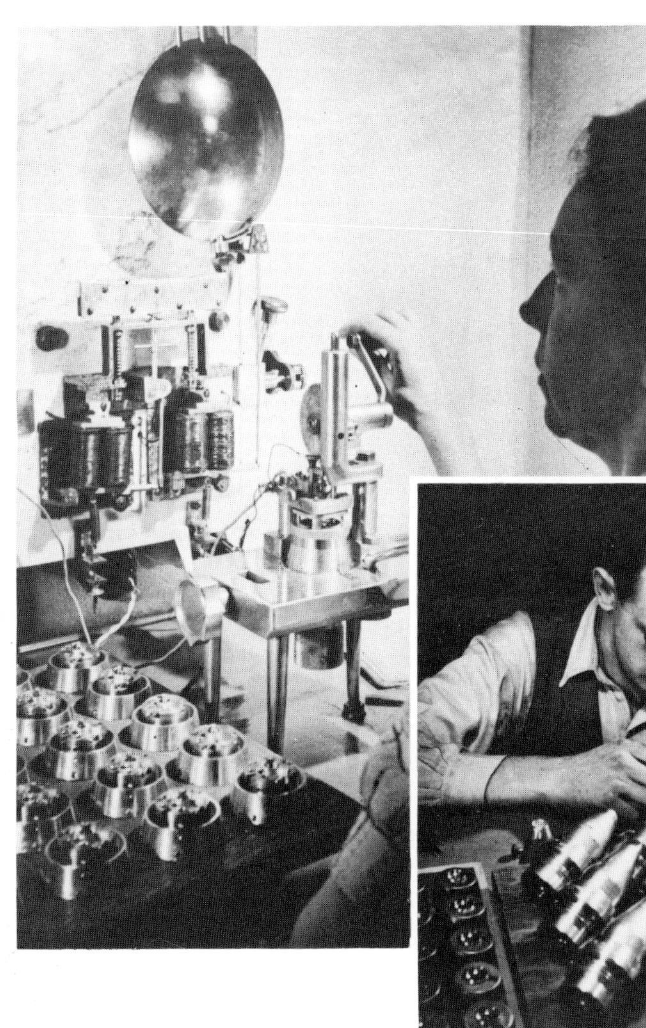

288

The manufacture and testing of mechanical time fuzes demands precision of a high order. Photo: Bofors A.G.

Anatomy of a Gun

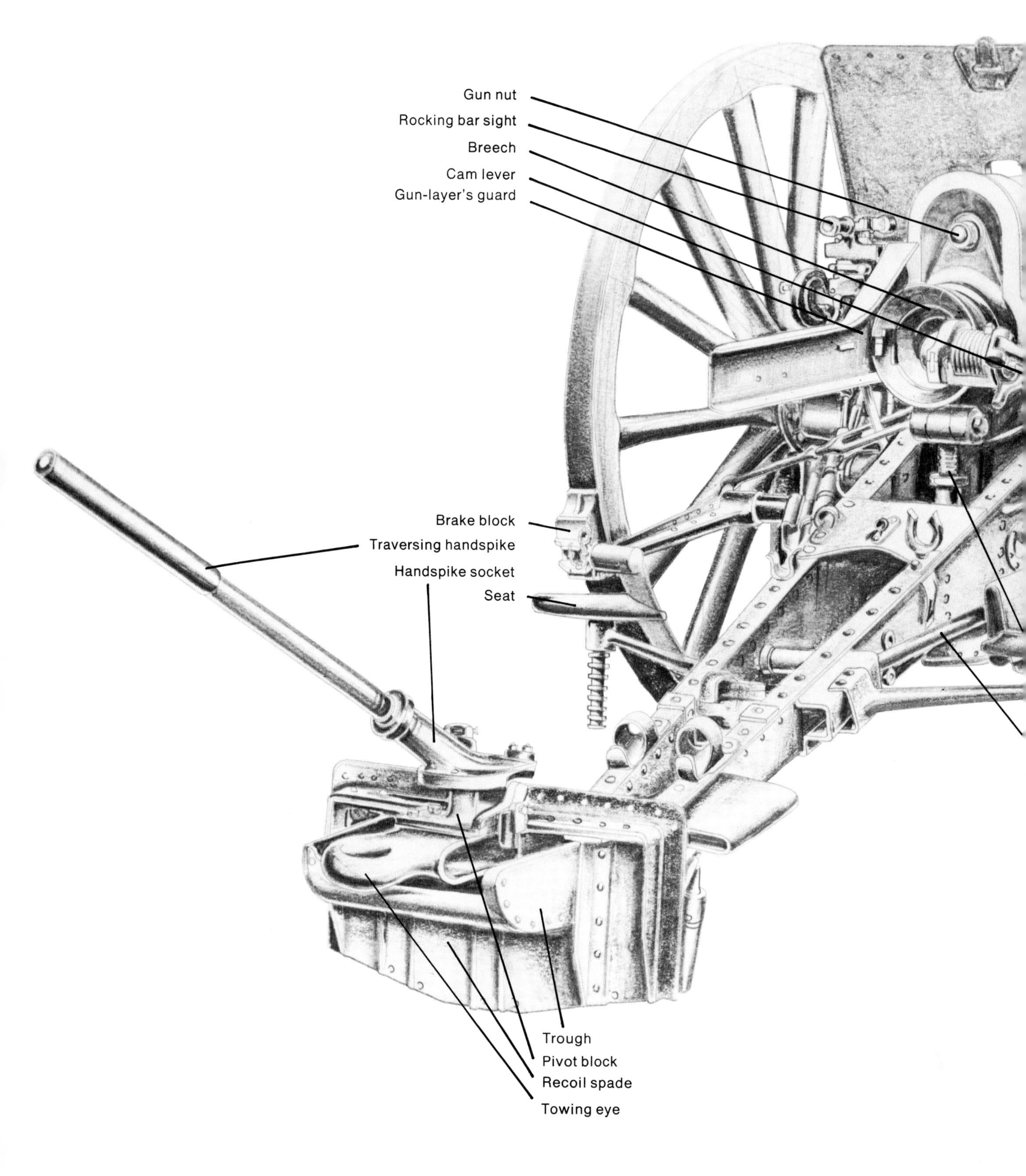

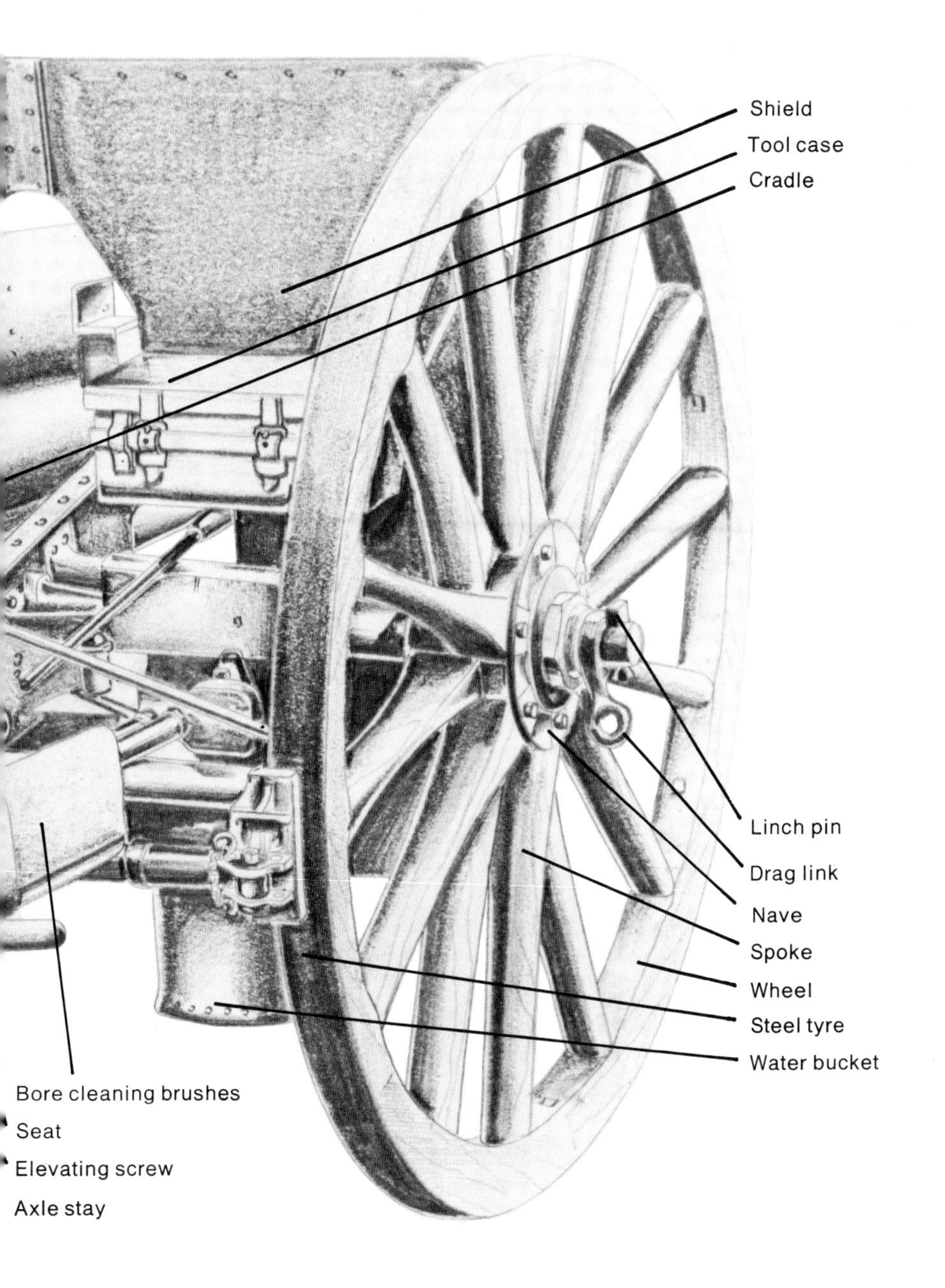

Ordnance, Breech-loading, converted, 15 pounder gun MK1 on carriage, field, B.L.C. 15 pounder MK 1 1907

Bibliography

Much of the information presented in this book came from hitherto unpublished sources, personal knowledge and a thousand-and-one minor notes taken over a period of many years. The following list is therefore not a list of references consulted during preparation of the book but is offered as a guide to those interested in further reading to obtain greater detail on the many aspects of artillery

Modern Guns and Gunnery Lt. Col. H. A. BETHELL *F. J. Cattermole London 1909*

Treatise on the Construction of Ordnance, 1879 *London HMSO 1879*

Treatise on Military Carriages 1911 *HMSO London 1911*

Artillery, Past Present and Future *Hadcock London 1894*

The Guns 1914-18 *Hogg Ballantine New York 1972*

The Guns 1939-45 *Hogg Ballantine New York 1970*

British Artillery & Ammunition of the First World War *Hogg & Thurston Ian Allen London 1972*

Coast Artillery Journal *Various copies*

Journal of the Royal Artillery *Various copies*

Proceedings of the Royal Artillery Institute *Various copies*

Proceedings of the Ordnance Select Committee *Various copies*

Picture index

Index